Apoptosis and Cancer

METHODS IN MOLECULAR BIOLOGY™

John M. Walker, SERIES EDITOR

METHODS IN MOLECULAR BIOLOGY™

Apoptosis and Cancer

Methods and Protocols

Edited by

Gil Mor

and

Ayesha B. Alvero

*Department of Obstetrics, Gynecology and Reproductive Sciences,
Yale University School of Medicine, New Haven, CT*

HUMANA PRESS ✳ TOTOWA, NEW JERSEY

Cover illustration: Cells in different stages of apoptosis. Supplied by Dr. Gil Mor.

Production Editor: Rhukea Hussain
Cover design by Karen Schulz

For additional copies, pricing for bulk purchases, and/or information about other Humana titles, contact Humana at the above address or at any of the following numbers: Tel.: 973-256-1699; Fax: 973-256-8341; E-mail: orders@humanapr.com; or visit our Website: www.humanapress.com

Printed in the United States of America. 10 9 8 7 6 5 4 3 2 1

e-ISBN 978-1-59745-339-4

Library of Congress Control Number: 2007935895

Preface

The aim of this book is to describe in clear detail the performance of contemporary techniques for studying the biology of apoptosis and its role in cancer.

Our understanding of the apoptotic pathway and its regulation has improved significantly in the past 15 years; however, as our knowledge of the intricate intracellular events increase, we realize the complexity of the regulatory mechanisms that control it.

Resistance to apoptosis has been defined as one of the hallmarks in tumor transformation, and more recently, it has become evident that it plays a major role in chemoresponse. Indeed, numerous studies have shown that defective apoptosis contributes to the acquisition of resistance to chemotherapy and immune effectors; therefore, the molecules regulating the apoptotic cascade become an excellent target for the development of new target-based therapies. This objective can only be achieved if appropriate tools for the evaluation and analysis of the apoptotic process are available.

This book is a collaboration between academic- and industry-based scientists. The chapters have the "technical development" characteristic of an academic environment and the "standardized system" necessary for translating a product to general use, as provided by the industry. Thus, the protocols described may aid an academic laboratory interested in further characterizing the mechanism of apoptosis and an industry laboratory, which is interested on identifying new target molecules or screening for new compounds with potential clinical use. We have tried to cover the newest available techniques as well as conventional basic techniques.

We thank all the authors for their outstanding contributions, and we are confident that the methods they described in this book will become an excellent tool in many laboratories and institutions.

Our sincere appreciation and gratefulness goes to Michele K. Montagna for her dedicated efforts in maintaining the organization of the manuscripts and correspondence between the editors and authors. She did an exceptional job of reminding authors and editors of the deadlines and of maintaining everything in a very organized manner. Special thanks to the Series Editor, Dr. John Walker, for his continued support and encouragement.

Finally, we hope this book will provide the necessary resources to everyone interested in the area of apoptosis, related not only to cancer biology but to other areas as well.

Gil Mor
Ayesha B. Alvero

Contents

Contributors

ALEXANDRU ALMASAN • *Department of Cancer Biology, Lerner Research Institute, and Department of Radiation Oncology, The Cleveland Clinic, Cleveland, OH*

MELCHOR ALVAREZ-MON • *CNB-CSIC R&D Associated Unit, Department of Medicine, University of Alcalá, Alcalá de Henares, Madrid, Spain; Research Unit, Industrial Farmacéutica Cantabria, Madrid, Spain; Immune System Diseases and Oncology Service, University Hospital "Príncipe de Asturias", Alcalá de Henares, Madrid, Spain*

AYESHA B. ALVERO • *Department of Obstetrics, Gynecology and Reproductive Sciences, Yale University School of Medicine, New Haven, CT*

MARTINE AUBERT • *Fred Hutchinson CRC, Seattle, WA*

HUGO BARCENILLA • *CNB-CSIC R&D Associated Unit, Department of Medicine, University of Alcalá, Alcalá de Henares, Madrid, Spain; Research Unit, Industrial Farmacéutica Cantabria, Madrid, Spain; Immune System Diseases and Oncology Service, University Hospital "Príncipe de Asturias", Alcalá de Henares, Madrid, Spain*

STACEY L. BROWER • *Precision Therapeutics, Pittsburgh, PA*

NICHOLAS M. BROWN • *Department of Pathology, University of Iowa College of Medicine, Iowa City, IA*

ROBERT F. BULLEIT • *Promega Corporation, Madison, WI*

JASON E. BUSH • *Precision Therapeutics, Pittsburgh, PA*

RUI CHEN • *Department of Obstetrics, Gynecology and Reproductive Sciences, Yale University School of Medicine, New Haven, CT*

JIN Q. CHENG • *H. Lee Moffitt Cancer Center and Research Institute, Tampa, FL*

JEAN-BERNARD DENAULT • *The Burnham Institute for Medical Research, La Jolla, CA*

DAVID DIAZ • *CNB-CSIC R&D Associated Unit, Department of Medicine, University of Alcalá, Alcalá de Henares, Madrid, Spain; Research Unit, Industrial Farmacéutica Cantabria, Madrid, Spain; Immune System Diseases and Oncology Service, University Hospital "Príncipe de Asturias", Alcalá de Henares, Madrid, Spain*

JEFFREY E. FENSTERER • *Precision Therapeutics, Pittsburgh, PA*

RICHARD FOX • *University of Washington, Seattle, WA*

HAN-HSUAN FU • *Department of Obstetrics, Gynecology and Reproductive Sciences, Yale University School of Medicine, New Haven, CT*

ROBERT J. GRIFFIN • *University of Minnesota Medical School, Minneapolis, MN*

TOBIAS L. HAAS • *Division of Apoptosis Regulation, German Cancer Research Center, Im Neuenheimer Feld, Heidelberg, Germany*

GARY L. JOHNSON • *Immunochemistry Technologies, LLC, Bloomington, MN*

HARRIET M. KLUGER • *Department of Medicine, Yale University School of Medicine, New Haven, CT*

C. MICHAEL KNUDSON • *Department of Pathology, University of Iowa College of Medicine, Iowa City, IA*

BRIAN W. LEE • *Immunochemistry Technologies, LLC, Bloomington, MN*

SUPARNA MAZUMDER • *Department of Cancer Biology, Lerner Research Institute, The Cleveland Clinic, Cleveland, OH*

MARY M. MCCARTHY • *Department of Medicine, Yale University School of Medicine, New Haven, CT*

JORGE MONSERRAT • *CNB-CSIC R&D Associated Unit, Department of Medicine, University of Alcalá, Alcalá de Henares, Madrid, Spain; Research Unit, Industrial Farmacéutica Cantabria, Madrid, Spain; Immune System Diseases and Oncology Service, University Hospital "Príncipe de Asturias", Alcalá de Henares, Madrid, Spain*

MICHELE K. MONTAGNA • *Department of Obstetrics, Gynecology and Reproductive Sciences, Yale University School of Medicine, New Haven, CT*

GIL MOR • *Department of Obstetrics, Gynecology and Reproductive Sciences, Yale University School of Medicine, New Haven, CT*

JECHIEL MOR • *Department of Obstetrics, Gynecology and Reproductive Sciences, Yale University School of Medicine, New Haven, CT*

RICHARD A. MORAVEC • *Promega Corporation, Madison, WI*

ANDREW L. NILES • *Promega Corporation, Madison, WI*

MARTHA A. O'BRIEN • *Promega Corporation, Madison, WI*

MICHAEL R. OLIN • *University of Minnesota College of Veterinary Medicine, St. Paul, MN*

ELAH PICK • *Department of Medicine, Yale University School of Medicine, New Haven, CT*

DRAGOS PLESCA • *Department of Cancer Biology, Lerner Research Institute, The Cleveland Clinic, Cleveland, OH; School of Biomedical Sciences; Kent State University, Kent, OH*

ALFREDO PRIETO • *CNB-CSIC R&D Associated Unit, Department of Medicine, University of Alcalá, Alcalá de Henares, Madrid, Spain; Research Unit, Industrial Farmacéutica Cantabria, Madrid, Spain; Immune System Diseases and Oncology Service, University Hospital "Príncipe de Asturias", Alcalá de Henares, Madrid, Spain*

EDUARDO REYES • *CNB-CSIC R&D Associated Unit, Department of Medicine, University of Alcalá, Alcalá de Henares, Madrid, Spain; Research Unit, Industrial Farmacéutica Cantabria, Madrid, Spain; Immune System Diseases and Oncology Service, University Hospital "Príncipe de Asturias", Alcalá de Henares, Madrid, Spain*

TERRY L. RISS • *Promega Corporation, Madison, WI*

GUY S. SALVESEN • *Program in Cell Death and Apoptosis Research, The Burnham Institute for Medical Research, La Jolla, CA*

DAN-ARIN SILASI • *Department of Obstetrics, Gynecology and Reproductive Sciences, Yale University School of Medicine, New Haven, CT*

CHRISTOPHER J. VEGA • *Leica Microsystems, Bannockburn, IL*

HENNING WALCZAK • *Division of Apoptosis Regulation, German Cancer Research Center, Im Neuenheimer Feld, Heidelberg, Germany*

JIA-WANG WANG • *H. Lee Moffitt Cancer Center and Research Institute, Tampa, FL*

1

Modulation of Apoptosis to Reverse Chemoresistance

Gil Mor, Michele K. Montagna, and Ayesha B. Alvero

Summary

Interference with the innate apoptotic activity is a hallmark of neoplastic transformation and tumor formation. Modulation of the apoptotic cascade has been proposed as a new approach for the treatment of cancer. In this chapter, we discuss the role of apoptosis in ovarian cancer and the use of phenoxodiol as a model for the regulation of apoptosis and potential use as chemosensitizer for chemoresistant ovarian cancer cells.

Key Words: Apoptosis; ovarian cancer; X-linked inhibitor of apoptosis protein (XIAP); phenoxodiol; caspases.

1. Introduction

Ovarian cancer is the fourth leading cause of cancer-related deaths in women and is the most lethal of the gynecological malignancies *(1)*. One in 70 women will develop ovarian cancer and one out of 100 women will die from it. The high mortality rate is due in part to the lack of means to detect early disease such that approximately 80% of patients are initially diagnosed in advanced-staged disease. In these patients, 80–90% will initially respond to chemotherapy; however, less than 10–15% will remain in remission because of the subsequent development of chemoresistance. Treatment advances have led to improved 5-year survival, approaching 45%; however, no advances have been made in the overall survival. One way to improve survival may be by influencing the apoptotic pathways.

From: *Methods in Molecular Biology, vol. 414: Apoptosis and Cancer*
Edited by: G. Mor and A. B. Alvero © Humana Press Inc., Totowa, NJ

Chemotherapy in the treatment of cancer was introduced into the clinical practice more than 50 years ago. Although this form of therapy has been successful in the treatment of some forms of cancer, it has not been the case for the majority of epithelial tumors of the breast, colon, lung, and ovary. Initially, the development of chemotherapeutic agents was based on the observation that tumor cells proliferate faster than normal cells. Therefore, the original strategy was to interfere with DNA replication or cellular metabolism. A better understanding of the molecular mechanisms of apoptosis and its function in normal physiology has resulted in a better understanding of the effect of chemotherapy and the mechanisms of chemoresistance. Current understanding suggests that the induction of apoptosis in target cells is a key mechanism for most anti-tumor therapies, including chemotherapy, γ-radiation, immunotherapy, and cytokines *(2)*. Therefore, *defects in apoptosis may cause resistance.* The realization that apoptosis is a key factor that contributes to the anti-tumor activity of chemotherapeutic drugs has allowed us to understand how drug resistance may arise and to look for new approaches for the treatment of cancer.

2. The Apoptotic Cascade

Apoptosis is characterized by morphological changes including cell shrinkage, membrane blebbing, chromatin condensation, and nuclear fragmentation *(3)*. All these changes are the result of the activation of a cascade of intracellular factors known as caspases. Caspases are highly specific proteases synthesized as zymogens and activated by cleavage at aspartate, which generates the large and small subunits of the mature enzyme *(3,4)*. These aspartate cleavage sites are themselves caspase sites. Therefore, caspases can collaborate in the proteolytic cascade by activating themselves and each other *(4,5)*.

Within these cascades, caspases can be divided into "initiator" caspases and downstream "effectors" of apoptosis. Initiator caspases, such as caspase-8 and caspase-9, mediate their oligomerization and auto-activation in response to specific upstream signals. The best-documented pathway of caspase activation is the assembly of the death-induced signaling complex (DISC) induced by the binding of the members of the death receptor family (Fas, TNF-related apoptosis inducing ligand (TRAIL), and tumor necrosis factor (TNFR1)) to its ligand (reviewed in **refs 6–10**). Caspase-8 is recruited to the DISC, undergoes spontaneous auto-activation, and activates downstream caspase-3 *(11)* (*see* **Fig. 1**). Another well-described pathway is the mitochondrial pathway, which is initiated by the release of cytochrome c from the mitochondria to the cytosol, where it assembles with Apaf-1 and the caspase-9 holoenzyme, leading

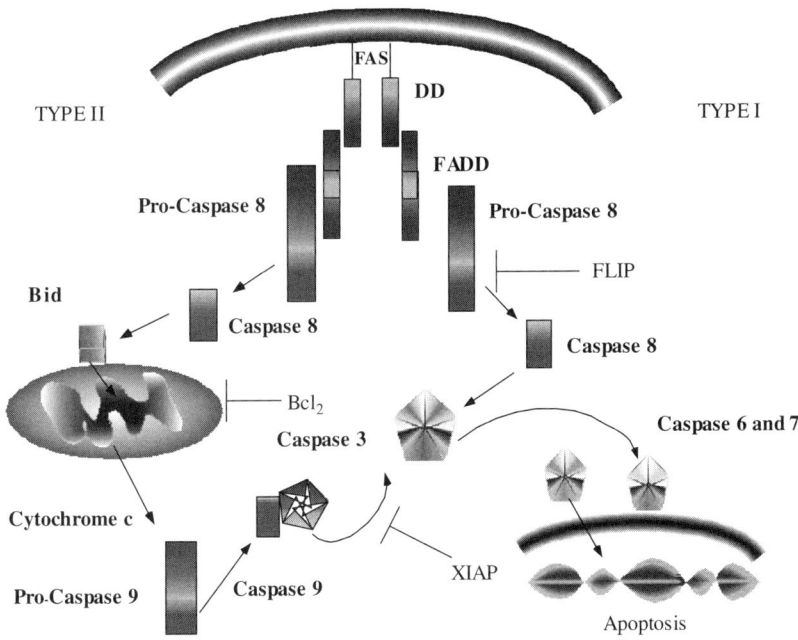

Fig. 1. The apoptotic cascade. Type I represents the death receptor pathway mediated by caspase-8 and Type II represents the mitochondrial pathway regulated by Bcl2 family members.

to the activation of caspase-9, which also activates downstream caspase-3 (reviewed in **refs** *4,12–14*). In contrast to caspase-8 and caspase-9, there are conflicting depictions for the role of caspase-2. Caspase-2 has been shown to be an important initiator caspase of the mitochondrial pathway, acting upstream of caspase-9 *(15–17)*. However, other studies show that caspase-2 activation occurs downstream of the mitochondria and caspase-9, and therefore, caspase-2 is an effector and not an initiator caspase *(18)*. Thus, the placement of caspase-2 in the apoptotic cascade is controversial. In addition, its mode of activation and regulation has not been characterized. The limited data available on caspase-2 are brought in part by earlier findings showing the absence of severe phenotype in caspase-2-deficient mice *(19)*, leading to the speculation that it does not play a significant role in apoptosis. Recent studies, however, demonstrated caspase-2's central role in cytotoxic stress-induced apoptosis in some human cell lines *(20,21)*. In addition, our preliminary data in epithelial ovarian cancer cells (EOC) cells show that caspase-2 is one of the earliest caspase activated in response to chemotherapy and that it is essential for the induction of the apoptotic cascade. Moreover, we have identified caspase-2 activation as one of the steps occurring in chemosensitive EOC cells in response to chemotherapy, but not in chemoresistant cells. This suggests that alterations in the regulation of caspase-2 may be one of the ways EOC cells develop chemoresistance.

The effector caspases include caspase-3, caspase-6, and caspase-7, which cleave cellular substrates and precipitate apoptotic death. To date, more than 280 caspase targets have been identified *(22)*. These include multiple proteins involved in cell adhesion, cell cycle regulation, DNA synthesis, cleavage and repair, RNA synthesis and splicing, protein synthesis, and those that make up the cytoskeletal and nuclear structures. Our preliminary results show that two of these proteins, vimentin and keratin, are differentially expressed in cells responding to therapy and in those that are not responding. Some of these proteins may "leak" to the circulation and be detected in the serum. Thus, one of our hypotheses is that downstream targets of the apoptotic cascade are potential markers of chemoresponse that warrant further investigation. Currently, the best marker available in the clinics to detect chemoresponse is cancer antigen 125 (CA125). In first-line treatment, CA125 has an established role in monitoring the efficacy of treatment *(23,24)*. Serial changes in its concentrations are well correlated with response and survival *(25)*. However, in second-line treatment, where chemoresistance is widely seen, several studies identify patients in whom CA125 levels are in discordance with the radiographic evaluation of changes in tumor load *(26–28)*. The identification of sensitive and specific markers of chemoresponse will provide a way to immediately monitor response to treatment and will aid in the tailoring of therapy in ovarian cancer patients, especially those with recurrent disease and in their second line of treatment.

3. Inhibitors of Apoptosis and Chemoresistance

Each step in the apoptotic cascade is delicately controlled by intracellular factors that can block the apoptotic pathway either at the "initiator" or "effector" level. There are three groups of proteins that are known to inhibit apoptosis (*see* **Fig. 1**): (i) the bcl2 family of proteins, which stabilize the mitochondrial membranes and prevent cytochrome c release (reviewed in **refs** *12,29,30*); (ii) the FADD-like interleukin-1 converting enzyme (FLICE)-inhibitory proteins (FLIP), which are caspase-8-like proteins and interfere with caspase-8 for binding to the DISC, thus preventing caspase-8 oligomerization and auto-activation (reviewed in **refs** *31–33*); and (iii) the inhibitors of apoptosis proteins (IAP), which constitute a family of evolutionarily conserved apoptotic suppressors (reviewed in **refs** *34–36*). All these inhibitory factors are, in the normal cell, inactivated in response to apoptotic signals such as hormone withdrawal, DNA damage, or activation of death receptors. In chemoresistant EOC cells, however, these blockers have been shown to remain active even after drug treatment. XIAP, which is the prototype of the IAP family of proteins, is a unique regulatory protein because of its capacity to block the apoptotic cascade both at the initiation and effector levels. It is capable of binding to pro-caspase-9, thus preventing the activation of the mitochondrial pathway, and it also binds and inhibits the effector caspases, caspase-3 and caspase-7 *(37)* (*see* **Fig. 1**). Several studies, including our own, have shown that the intracellular blockers of apoptosis may contribute to drug resistance *(34,35,38)*. Indeed, we have previously shown that XIAP is highly expressed in all tested ovarian cancer samples *(39)*.

4. Regulation of XIAP

The role of XIAP in chemoresistance is already well established. Numerous studies have demonstrated that its up-regulation confers resistance to chemotherapy and its down-regulation confers sensitivity *(34,40–43)*. The mechanisms of XIAP regulation, which ultimately leads to its inactivation and hence apoptosis in sensitive cells, are, however, not clearly understood. Similarly, the mechanisms that prevent its inactivation in resistant cells are unknown. Three negative regulators of XIAP have been identified: (i) XIAP-associated factor 1 (XAF-1) is a nuclear protein that binds XIAP and antagonizes its ability to suppress the caspases *(44,45)*; (ii) Smac/direct inhibitor of apoptosis-binding protein with low pI (DIABLO) is a mitochondrial protein that also binds and inhibits XIAP *(46,47)*; and (iii) Omi/HtrA2 is another mitochondrial protein that can bind XIAP and in addition, through its protease activity, can induce XIAP cleavage and inactivation *(48–52)*.

Recently, the relationship between XIAP and Akt has been characterized. Dan et al. *(53)* showed that XIAP regulates Akt activity in response to cisplatin. Specifically, they showed that down-regulation of XIAP by cisplatin induces Akt cleavage and that XIAP over-expression induces Akt phosphorylation. Another study demonstrated that XIAP is a substrate of Akt and that phosphorylation of XIAP by Akt stabilizes XIAP and prevents its degradation by the ubiquitin-proteasome pathway *(54)*.

Although all of these individual mechanisms may be significant, the mode of regulation and the sequence of events of XIAP inactivation, which is responsible for chemosensitivity and hence is possibly altered in chemoresistant cells, remain to be identified and characterized. We hypothesize that the cleavage of XIAP by Omi/HtrA2 is required for the full activation of the apoptotic cascade as it results in the formation of the p30 XIAP-cleaved product that can act as a dominant negative of XIAP. We have characterized the effect of phenoxodiol on a panel of EOC cell lines isolated from ascites *(55)*. The majority of these cells exhibit in vitro resistance to a wide range of chemotherapeutic agents. Our initial studies compared the effect of phenoxodiol on these chemoresistant cells and on normal ovarian surface epithelial cells. We found that phenoxodiol was highly effective in inducing cell death in the cancer cell lines but not in normal cells *(56)*. When we compared the effect of phenoxodiol and genistein on cell viability, phenoxodiol was found to be 30 times more effective than genistein. In CP70 cells, the IC_{50} for phenoxodiol was 1.35 μM, whereas the IC_{50} for genistein was 38.95 μM.

The most remarkable effect of phenoxodiol was observed in EOC cells that are resistant to carboplatin and paclitaxel. Treatment with phenoxodiol for 24 or 48 h induced a significant increase in cell death in cells where carbo-platin and paclitaxel were ineffective. In vivo studies using a mouse xenograft model confirmed the in vitro observation. When phenoxodiol was dosed i.p. at 20 mg/kg every day for 6 days, the optimal treatment/control (T/C) was 24.7% ($p < 0.02$) indicating that phenoxodiol was effective at inhibiting tumor growth *(56,57)*. Furthermore, there was a 3.1-fold reduction in terminal tumor mass in phenoxodiol-treated groups compared with control *(56)*. An important finding was the lack of toxicity, because no toxic side effects were noted at this dose and animals continued to gain weight throughout the therapeutic window.

5. Phenoxodiol Anti-cancer Mechanism of Action

The induction of apoptosis in target cells is thought to be a key mechanism for most anti-tumor therapies, including chemotherapy, γ-radiation, and immunotherapy *(2)*. Therefore, defects in apoptosis may cause chemoresistance. The realization that apoptosis is a key factor that contributes to the anti-tumor activity of chemotherapeutic drugs has allowed us to understand how drug resistance may arise and to look for new treatment approaches. Several studies, including ours, showed that the intracellular blockers of apoptosis are the main players in drug resistance *(34,35,38)*. Indeed, we previously showed that XIAP is highly expressed in ovarian cancer.

Our studies with chemoresistant EOC cells showed that phenoxodiol is able to fully activate the apoptotic pathway, as evidenced by the activation of the main effector caspase, caspase-3, in cells that did not undergo apoptosis after treatment with carboplatin, paclitaxel, gemcitabine, or docetaxel *(56,57)*. This suggests that the apoptotic pathway is functional in these cells but is activated only in response to phenoxodiol. To understand the effect of phenoxodiol on chemoresistant EOC cells, we evaluated changes in the expression and function of several proteins regulating the apoptotic pathway. One of the earliest effects we observed in EOC cells following phenoxodiol treatment was the activation of caspase-2, which occurred 4 h post-treatment. Caspase-2 is an initiator caspase that acts upstream of the mitochondria, possibly through Bid, and promotes cytochrome c release and apoptosis. Its activation has been associated with the stress-induced cell death, which includes anti-Fas, cytokine deprivation, β-amyloid, etoposide, chemotherapeutic agents, and other stress stimuli *(16,17)*.

In addition to caspase-2 activation, Bid activation and XIAP down-regulation were also observed *(57)*. Further tests showed that Bid activation is secondary to caspase-2 activation and that XIAP down-regulation was a result of proteasome degradation *(57)*. Moreover, we showed that in EOC cells, phenoxodiol treatment induces the down-regulation of both phosphorylated Akt and total Akt *(57)*. Taken together, these results suggest that in chemoresistant EOC cells, caspase-2 activation is essential for the induction of apoptosis. Thus, alterations in the regulation of caspase-2 may be one of the ways through which EOC cells develop chemoresistance.

The down-regulation of XIAP is another early event that occurs after phenoxodiol treatment. We observed a decrease in p45 XIAP at 4 h post-treatment and the appearance of the p30-cleaved form 16 h post-treatment. In contrast to the decrease in p45 XIAP, which was proteasome dependent, cleavage of XIAP and the appearance of the p30 fragment were shown to be caspase dependent *(57)*. Thus, in addition to caspase-2 activation, the early down-regulation of XIAP may be another requirement for chemoresistant EOC cells to undergo apoptosis. XIAP down-regulation, concomitant with caspase-2 activation, probably allows the full activation of the caspase cascade. Thus, by simultaneously activating an initiator caspase and down-regulating a potent caspase inhibitor, phenoxodiol is able to induce apoptosis in EOC cells. However, the initial target for phenoxodiol, which triggers this cascade of events, is not clearly defined.

A potential candidate is ceramide. Ceramide has been shown to be one of the upstream signals activating caspase-2, which then can activate caspase-8 and regulators of mitochondria integrity *(20)*. The regulation of sphingosine kinase (SK), ceramide, and sphingosine-1 phosphate (S1P) has an integral role in the maintenance of the balance between cell death and survival. Cancer cells are characterized by overexpression of SK and increased production of S1P. S1P then initiates a series of events that enhance cell proliferation and inhibit pro-apoptotic signals. One of the pathways activated by S1P is the Akt survival pathway *(58)*, which is highly expressed in ovarian cancer and has been related to chemoresistance *(53,57,59)*.

The levels of pAkt in ovarian cancer cells are significantly inhibited by phenoxodiol treatment, and this effect precedes the activation of members of the mitochondrial pathway, especially Bid and Bax *(57)*. Akt has been shown to inhibit Bax and caspase-9 by phosphorylation, therefore we hypothesized that the inhibitory effect of phenoxodiol on Akt may release the pro-apoptotic proteins that are under Akt control and therefore re-activate the apoptotic cascade. This hypothesis is supported by the fact that the decrease in pAkt is followed by Bax, caspase-8, caspase-9, and caspase-3 activation *(57,60)*.

Other mechanisms for phenoxodiol have been described. Aguero et al. *(61)* recently showed that phenoxodiol promotes G1 arrest by the loss of cyclin-dependent kinase 2 activity secondary to p53-independent p21[WAF1] induction. In addition, Gamble et al. *(62)* showed that phenoxodiol had anti-angiogenic properties. It is able to inhibit endothelial cell proliferation, migration, and capillary tube formation. In addition, it is also able to inhibit the expression of a major matrix-degrading enzyme, matrix metalloproteinase 2. Taken together, these findings show that in cancer cells, phenoxodiol is able to activate the apoptotic cascade, induce cell cycle arrest, and inhibit angiogenesis.

6. Phenoxodiol as a Chemosensitizer

One of the remarkable effects of phenoxodiol in vitro and in animal models is its capacity to reverse resistance. Trophoblast cells are resistant to Fas-mediated apoptosis. This resistance is due to the expression of the anti-apoptotic protein FLIP, which inhibits caspase-8 activation at the DISC *(63–65)*. However, when trophoblast cells were pre-treated with phenoxodiol prior to Fas ligation, significant cell death was observed *(56,60)*. Chemosensitization to Fas was shown to be secondary to phenoxodiol-induced decrease in FLIP expression, which allowed caspase-8 activation following Fas ligation by FasL *(56,60)*. EOC cells are also resistant to Fas-mediated apoptosis. Suboptimal exposure to phenoxodiol prior to Fas ligation also sensitized these cells to Fas-mediated apoptosis *(56)*. In addition, phenoxodiol is able to sensitize chemoresistant EOC cells. The IC_{50} for carboplatin, paclitaxel, gemcitabine, and docetaxel significantly decrease when EOC cells were pre-treated with phenoxodiol *(57,66)*. In animal models, treatment with phenoxodiol sensitized to cisplatin, paclitaxel, and gemcitabine *(56,57,67)*.

7. Conclusion

Targeting the apoptotic cascade represents a novel approach for the treatment of cancer and to potentially overcome chemoresistance. A better understanding of the mechanisms by which cancer cells control this cascade can be achieved by developing better methods to study apoptosis. The chapters in this book describe new tools that can facilitate these studies.

References

1. Schwartz, P. E. (2002) *Cancer Treat Res* **107**, 99–118.
2. Kaufmann, S. H., and Earnshaw, W. C. (2000) *Exp Cell Res* **256**, 42–49.
3. Wyllie, A. H., Kerr, J. F., and Currie, A. R. (1980) *Int Rev Cytol* **68**, 251–306.
4. Cain, K., Bratton, S. B., and Cohen, G. M. (2002) *Biochimie* **84**, 203–214.
5. Cohen, G. M. (1997) *Biochem J* **326** (Pt. 1), 1–16.
6. Reichmann, E. (2002) *Semin Cancer Biol* **12**, 309–315.
7. Ozoren, N., and El-Deiry, W. S. (2003) *Semin Cancer Biol* **13**, 135–147.
8. Dempsey, P. W., Doyle, S. E., He, J. Q., and Cheng, G. (2003) *Cytokine Growth Factor Rev* **14**, 193–209.
9. Gaur, U., and Aggarwal, B. B. (2003) *Biochem Pharmacol* **66**, 1403–1408.
10. Ashkenazi, A., and Dixit, V. M. (1998) *Science* **281**, 1305–1308.
11. Muzio, M., Stockwell, B. R., Stennicke, H. R., Salvesen, G. S., and Dixit, V. M. (1998) *J Biol Chem* **273**, 2926–2930.
12. Daniel, P. T., Schulze-Osthoff, K., Belka, C., and Guner, D. (2003) *Essays Biochem* **39**, 73–88.
13. Green, D. R., and Kroemer, G. (2004) *Science* **305**, 626–629.
14. Nicholson, D. W., and Thornberry, N. A. (2003) *Science* **299**, 214–215.
15. Lassus, P., Opitz-Araya, X., and Lazebnik, Y. (2002) *Science* **297**, 1352–1354.
16. Guo, Y., Srinivasula, S. M., Druilhe, A., Fernandes-Alnemri, T., and Alnemri, E. S. (2002) *J Biol Chem* **277**, 13430–13437.
17. Robertson, J. D., Enoksson, M., Suomela, M., Zhivotovsky, B., and Orrenius, S. (2002) *J Biol Chem* **277**, 29803–29809.
18. O'Reilly, L. A., Ekert, P., Harvey, N., Marsden, V., Cullen, L., Vaux, D. L., Hacker, G., Magnusson, C., Pakusch, M., Cecconi, F., et al. (2002) *Cell Death Differ* **9**, 832–841.
19. Bergeron, L., Perez, G. I., Macdonald, G., Shi, L., Sun, Y., Jurisicova, A., Varmuza, S., Latham, K. E., Flaws, J. A., Salter, J. C., et al. (1998) *Genes Dev* **12**, 1304–1314.
20. Lin, C. F., Chen, C. L., Chang, W. T., Jan, M. S., Hsu, L. J., Wu, R. H., Tang, M. J., Chang, W. C., and Lin, Y. S. (2004) *J Biol Chem* **279**, 40755–40761.
21. Wagner, K. W., Engels, I. H., and Deveraux, Q. L. (2004) *J Biol Chem* **279**, 35047–35052.
22. Fischer, U., Janicke, R. U., and Schulze-Osthoff, K. (2003) *Cell Death Differ* **10**, 76–100.
23. Meyer, T., and Rustin, G. J. (2000) *Br J Cancer* **82**, 1535–1538.
24. Berek, J. S., Bertelsen, K., du Bois, A., Brady, M. F., Carmichael, J., Eisenhauer, E. A., Gore, M., Grenman, S., Hamilton, T. C., Hansen, S. W., et al. (1999) *Ann Oncol* **10** (Suppl 1), 87–92.

25. Rustin, G. J., Nelstrop, A. E., Crawford, M., Ledermann, J., Lambert, H. E., Coleman, R., Johnson, J., Evans, H., Brown, S., and Oster, W. (1997) *J Clin Oncol* **15**, 172–176.

26. Davelaar, E. M., Bonfrer, J. M., Verstraeten, R. A., ten Bokkel Huinink, W. W., and Kenemans, P. (1996) *Cancer* **78**, 118–127.

27. Eisenhauer, E. A., ten Bokkel Huinink, W. W., Swenerton, K. D., Gianni, L., Myles, J., van der Burg, M. E., Kerr, I., Vermorken, J. B., Buser, K., Colombo, N., et al. (1994) *J Clin Oncol* **12**, 2654–2666.

28. Morgan, R. J., Jr., Speyer, J., Doroshow, J. H., Margolin, K., Raschko, J., Sorich, J., Akman, S., Leong, L., Somlo, G., Vasilev, S., et al. (1995) *Gynecol Oncol* **58**, 79–85.

29. Willis, S., Day, C. L., Hinds, M. G., and Huang, D. C. (2003) *J Cell Sci* **116**, 4053–4056.

30. Cory, S., and Adams, J. M. (2002) *Nat Rev Cancer* **2**, 647–656.

31. Krueger, A., Baumann, S., Krammer, P. H., and Kirchhoff, S. (2001) *Mol Cell Biol* **21**, 8247–8254.

32. Thome, M., and Tschopp, J. (2001) *Nat Rev Immunol* **1**, 50–58.

33. Tschopp, J., Irmler, M., and Thome, M. (1998) *Curr Opin Immunol* **10**, 552–558.

34. Cheng, J. Q., Jiang, X., Fraser, M., Li, M., Dan, H. C., Sun, M., and Tsang, B. K. (2002) *Drug Resist Update* **5**, 131–146.

35. Li, J., Sasaki, H., Sheng, Y. L., Schneiderman, D., Xiao, C. W., Kotsuji, F., and Tsang, B. K. (2000) *Biol Signals Recept* **9**, 122–130.

36. Yang, Y. L., and Li, X. M. (2000) *Cell Res* **10**, 169–177.

37. Deveraux, Q. L., Takahashi, R., Salvesen, G. S., and Reed, J. C. (1997) *Nature* **388**, 300–304.

38. Reed, J. C. (2002) *Nat Rev Drug Discov* **1**, 111–121.

39. Kamsteeg, M., Rutherford, T., Sapi, E., Hanczaruk, B., Shahabi, S., Flick, M., Brown, D., and Mor, G. (2003) *Oncogene* **22**, 2611–2620.

40. Fraser, M., Leung, B. M., Yan, X., Dan, H. C., Cheng, J. Q., and Tsang, B. K. (2003) *Cancer Res* **63**, 7081–7088.

41. Mansouri, A., Zhang, Q., Ridgway, L. D., Tian, L., and Claret, F. X. (2003) *Oncol Res* **13**, 399–404.

42. Sapi, E., Alvero, A. B., Chen, W., O'Malley, D., Hao, X. Y., Dwipoyono, B., Garg, M., Kamsteeg, M., Rutherford, T., and Mor, G. (2004) *Oncol Res* **14**, 567–578.

43. Sasaki, H., Sheng, Y., Kotsuji, F., and Tsang, B. K. (2000) *Cancer Res* **60**, 5659–5666.

44. Fong, W. G., Liston, P., Rajcan-Separovic, E., St Jean, M., Craig, C., and Korneluk, R. G. (2000) *Genomics* **70**, 113–122.

45. Liston, P., Fong, W. G., Kelly, N. L., Toji, S., Miyazaki, T., Conte, D., Tamai, K., Craig, C. G., McBurney, M. W., and Korneluk, R. G. (2001) *Nat Cell Biol* **3**, 128–133.

46. Verhagen, A. M., Ekert, P. G., Pakusch, M., Silke, J., Connolly, L. M., Reid, G. E., Moritz, R. L., Simpson, R. J., and Vaux, D. L. (2000) *Cell* **102**, 43–53.

47. Du, C., Fang, M., Li, Y., Li, L., and Wang, X. (2000) *Cell* **102**, 33–42.

48. Suzuki, Y., Imai, Y., Nakayama, H., Takahashi, K., Takio, K., and Takahashi, R. (2001) *Mol Cell* **8**, 613–21.

49. Yang, Q. H., Church-Hajduk, R., Ren, J., Newton, M. L., and Du, C. (2003) *Genes Dev* **17**, 1487–1496.

50. van Loo, G., van Gurp, M., Depuydt, B., Srinivasula, S. M., Rodriguez, I., Alnemri, E. S., Gevaert, K., Vandekerckhove, J., Declercq, W., and Vandenabeele, P. (2002) *Cell Death Differ* **9**, 20–26.

51. Verhagen, A. M., Silke, J., Ekert, P. G., Pakusch, M., Kaufmann, H., Connolly, L. M., Day, C. L., Tikoo, A., Burke, R., Wrobel, C., et al. (2002) *J Biol Chem* **277**, 445–454.

52. Hegde, R., Srinivasula, S. M., Zhang, Z., Wassell, R., Mukattash, R., Cilenti, L., DuBois, G., Lazebnik, Y., Zervos, A. S., Fernandes-Alnemri, T., et al. (2002) *J Biol Chem* **277**, 432–438.

53. Dan, H. C., Sun, M., Kaneko, S., Feldman, R. I., Nicosia, S. V., Wang, H. G., Tsang, B. K., and Cheng, J. Q. (2004) *J Biol Chem* **279**, 5405–5412.

54. Asselin, E., Mills, G. B., and Tsang, B. K. (2001) *Cancer Res* **61**, 1862–1868.

55. Flick, M. B., O'Malley, D., Rutherford, T., Rodov, S., Kamsteeg, M., Hao, X. Y., Schwartz, P. E., Kacinski, B. M., and Mor, G. (2004) *J Soc Gynecol Invest* **11**, 252–259.

56. Kamsteeg, M., Rutherford, T., Sapi, E., Hanczaruk, B., Shahabi, S., Flick, M., Brown, D., and Mor, G. (2003) *Oncogene* **22**, 2611–2620.

57. Alvero, A. B., O'Malley, D., Brown, D., Kelly, G., Garg, M., Chen, W., Rutherford, T., and Mor, G. (2006) *Cancer* **106**, 599–608.

58. Baudhuin, L. M., Jiang, Y., Zaslavsky, A., Ishii, I., Chun, J., and Xu, Y. (2004) *FASEB J* **18**, 341–343.

59. Dan, H. C., Jiang, K., Coppola, D., Hamilton, A., Nicosia, S. V., Sebti, S. M., and Cheng, J. Q. (2004) *Oncogene* **23**, 706–715.

60. Straszewski-Chavez, S. L., Abrahams, V. M., Funai, E. F., and Mor, G. (2004) *Mol Hum Reprod* **10**, 33–41.

61. Aguero, M. F., Facchinetti, M. M., Sheleg, Z., and Senderowicz, A. M. (2005) *Cancer Res* **65**, 3364–3373.

62. Gamble, J. R., Xia, P., Hahn, C. N., Drew, J. J., Drogemuller, C. J., Brown, D., and Vadas, M. A. (2006) *Int J Cancer* **118**, 2412–2420.

63. Straszewski-Chavez, S. L., Abrahams, V. M., and Mor, G. (2005) *Endocr Rev.* **26** (7):877–97.

64. Kataoka, T., Budd, R. C., Holler, N., Thome, M., Martinon, F., Irmler, M., Burns, K., Hahne, M., Kennedy, N., Kovacsovics, M., and Tschopp, J. (2000) *Curr Biol* **10**, 640–648.

65. Tschopp, J., Irmler, M., and Thome, M. (1998) *Curr Opin Immunol* **10**, 552–558.

66. Sapi, E., Chen, W., O'Malley, D., Hao, X., Dwipoyono, B., Garg, M., Kamsteeg, M., Rutherford, T., and Mor, G. (2004) *Anti-Cancer Drugs* **14**, 567–578.
67. Brown, D. M., Kelly, G. E., and Husband, A. J. (2005) *Mol Biotechnol* **30**, 253–270.

2

Caspase-3 Activation is a Critical Determinant of Genotoxic Stress-Induced Apoptosis

Suparna Mazumder, Dragos Plesca, and Alexandru Almasan

Summary

A number of methods have been developed to identify the cells that undergo apoptosis by analyzing the morphological, biochemical, and molecular changes that take place during this universal biological process. The best recognized biochemical hallmark of both early and late stages of apoptosis is the activation of cysteine proteases (caspases). Detection of active caspase-3 in cells and tissues is an important method for apoptosis induced by a wide variety of apoptotic signals. Most common assays for examining caspase-3 activation include immunostaining, immunoblotting for active caspase-3, colorimetric assays using fluorochrome substrates, as well as employing the fluorescein-labeled CaspaTag pan-caspase in situ detection kit.

Key Words: Caspase-3; apoptosis; γ-irradiation; PARP-1; flow cytometry; immunohistochemistry.

1. Introduction

Apoptosis, an evolutionary conserved genetic program of cell death in higher eukaryotes, is a basic process involved in cellular development and differentiation (*1,2*). Apoptosis may be essential for the prevention of tumor formation, and its deregulation is widely believed to be involved in pathogenesis of many diseases, including cancer (*3*). In almost all instances, deregulated cell proliferation and suppressed cell death together provide the underlying platform for neoplastic progression (*4*).

From: *Methods in Molecular Biology, vol. 414: Apoptosis and Cancer*
Edited by: G. Mor and A. B. Alvero © Humana Press Inc., Totowa, NJ

A critical process in apoptosis is the activation of a cascade of ICE/CED-3 family of cysteine proteases, termed caspases *(5,6)*. Caspases are intracellular cysteine proteases that mediate cell death and inflammation. Caspases have been originally identified in *Caenorhabditis elegans*. Later, mammalian homologs of these caspases have been discovered. Mammalian caspases, 14 members discovered to date, play distinct roles in apoptosis and inflammation. Specifically, caspase-3 is a major mediator of both apoptotic and necrotic cell death. Caspases are synthesized as inactive precursors or zymogens, which are activated by proteolytic cleavage to generate active enzymes that then may further proteolytically cleave other caspases or cellular proteins *(7)*. An active caspase consists of two large and two small subunits that form two heterodimers that associate in a tetramer *(8–10)*. They recognize a 4–5 amino acid sequence on the substrate, which has an aspartic acid residue at P4 position as a critical requirement. This residue, that is the target for specific cleavage, occurs at the carbonyl end of the aspartic acid residue *(11)*. Active caspase-9 cleaves procaspase-3, which then is required for many of the characteristic apoptotic nuclear changes. Downstream effector caspases, such as caspase-3, cleave and inactivate proteins crucial for the maintenance of cellular cytoskeleton, DNA repair, signal transduction, and cell-cycle control *(12)*. There are over 300 in vivo caspase substrates; among them are poly (ADP-ribose) polymerase (PARP-1) and ICAD/DFF45, the cleavage of which results in the liberation of a caspase-activated deoxyribonuclease (CAD) that is responsible for the oligonucleosome-size DNA fragmentation that is characteristic to most apoptotic cells *(13)*. The activation cascade of ICE/CED-3 family of caspases is a common and critical step in the execution phase of apoptosis, triggered by different factors, including genotoxic agents (e.g., γ-irradiation or treatment with anti neoplastic agents) *(5,14–16)*. Pharmacological inhibitors of caspase-3 can prevent the cell death following irradiation significantly indicating that caspase-3 activation is critical for genotoxic stress-induced apoptosis. Moreover, activation of caspase-3 can be an effective marker for the positive outcome of the different radio- and chemotherapeutic treatments of various cancers.

Caspase-3 can be detected through immunofluorescence and immunoblotting by using anti-active caspase-3 antibodies, by colorimetric assays employing fluorochrome substrates, and by flow cytometric methods, such as that using fluorochrome inhibitors of caspases (FLICAs). Activated caspases cleave many cellular proteins, and the resulting *signature* proteolytic fragments may also serve as useful markers. This chapter provides a few standard protocols that we have successfully used in our laboratory for a number of experimental systems, including cells grown in culture and as xenografts.

2. Materials

2.1. Immunocytochemistry for Detecting Active Caspase-3

1. Glass coverslips (22 mm × 22 mm) and slides (Fisher Scientific Co., IL, USA).
2. Formaldehyde: 4% in 1× PBS (dilute 37% formaldehyde stock in 1× PBS), make fresh each time.
3. Blocking buffer: 2% goat serum, 0.3% Triton X-100 in 1× PBS, sterile filtered.
4. Primary antibodies: anti-active caspase-3 (Cell Signaling Technology Inc., MA, USA).
5. Secondary antibody: fluorochrome-conjugated (Molecular Probes Invitrogen Corporation CA, USA).
6. Vectashield, mounting medium for fluorescence (Vector Laboratories Inc., CA, USA), with or without 4′,6′-diamidino-2-phenylindole hydrochloride (DAPI).
7. Nail polish.

2.2. Caspase-3 Activity Determination: Colorimetric Assay

1. Lysis buffer: 1% NP 40, 20 mM HEPES (pH 7.5), 4 mM EDTA. Just before use, add the following protease inhibitors: aprotinin (10 μg/ml), leupeptin (10 μg/ml), pepstatin (10 μg/ml), and phenyl methyl sulfonyl fluoride (PMSF) (1 mM).
2. Reaction buffer: 100 mM HEPES (pH 7.5), 20% v/v glycerol, 5 mM dithiothreitol (DTT), and 0.5 mM EDTA.
3. Caspase-3 substrate: Ac-DEVD-*p*-nitroanilide (Ac-DEVD-pNA) (Calbiochem, EMD Chemicals Inc., CA, USA), 20 mM stock in dimethyl sulfoxide (DMSO) (stable for >1 year at –20°C), 100 μM final concentration. Additional colorimetric as well as fluorometric substrates are available. The fluorometric substrates 7-amino-4-methylcoumarine (AMC) and 7-amino-4-trifluoromethylcoumarin (AFC) are more sensitive but require a fluorometer capable of detecting the 380/460 and 405/500 excitation/emission spectra, respectively. AFC can be also detected colorimetrically at 380 nM.
4. Microtiter plate reader, spectrophotometer, or fluorometer.

2.3. Detection of Caspase-3 Activation by Immunoblotting

1. 1× PBS.
2. Lysis buffer: 20 mM HEPES, pH 7.5, 1 mM EDTA, 150 mM NaCl, 1% NP-40, and 1 mM DTT with protease inhibitors (1 mM PMSF and 1 μg/ml leupeptin).
3. Bio-Rad Protein Assay reagent.
4. Bovine serum albumin (BSA).
5. Sodium dodecyl sulfate–polyacrylamide gel electrophoresis (SDS–PAGE).
6. Nitrocellulose membrane (e.g., Schleicher and Schull).
7. 1× PBST (1× PBS with 0.1% Tween 20).
8. Milk (non-fat dry milk).
9. Active caspase-3 (Cell Signaling) and PARP-1 (Cell Signaling) antibodies.

10. Secondary antibodies: anti-mouse or anti rabbit (Amersham, Biosciences, NJ, USA).
11. Chemiluminescent reagents: lumiglo (KPL Inc., MD, USA) or ECL (Amersham).
12. Disuccinimidyl suberate (DSS), final concentration 2 mM.
13. Conjugation buffer: 20 mM sodium phosphate, pH 7.5, containing 0.15 M NaCl, 20 mM HEPES, pH 7.0, and 100 mM carbonate/bicarbonate, pH 9.0.
14. Quenching buffer: 1 M Tris–HCl, pH 7.5.

2.4. Fluorescein-Labeled CaspaTag Pan-Caspase In Situ Assay Kit

2.4.1. Kit Components

1. FLICA reagent (FAM-VAD-FMK): for lyophilized vials, reconstitute one vial of lyophilized reagent with 50 µl DMSO and mix by swirling until completely dissolved (150× stock solution); working solution: 30×, dilute 1:5 in PBS, pH 7.4.
2. 10× wash buffer: 60 ml; working solution: dilute 1:10 in deionized water.
3. Fixative: 6 ml.
4. Propidium iodide (PI): 1 ml at 250 µg/ml.
5. Hoechst 33342 stain: 1 ml at 200 µg/ml.

2.4.2. Materials Not Supplied

1. Cultured cells with media.
2. 15-ml polystyrene centrifuge tubes.
3. Microscope slides.
4. Hemocytometer.
5. Centrifuge.
6. Vortexer.
7. PBS, pH 7.4.
8. DMSO.

3. Methods

Cells (2×10^5/ml) are irradiated at 4–20 Gy [^{137}Cs source, fixed dose rate of 2.8 Gy/min] *(14)* or treated with any of the DNA-damaging chemotherapeutic agents, such as the topoisomerase inhibitor etoposide (VP16, 10 µM) *(6,16)*.

3.1. Immunocytochemistry to Detect Active Caspase-3

Caspase-3, the major effector caspase, is one of the key executioners of apoptosis. In response to an apoptotic signal, cleavage of inactive caspase-3 occurs mainly at the Asp175 residue, and thereby, being activated. A specific antibody against active caspase-3 can be used to detect the apoptotic cells by immunocytochemistry.

1. For plating cells for this experiment, we use glass coverslips (sterilized by dipping in ethanol and passing through flame) that are placed into 6-well plates. Cells (1×10^5 cells/well) are seeded and grown overnight.

2. Remove media and rinse cells with $1\times$ PBS warmed to 37°C.
3. To fix the cells, add 1–2 ml 4% formaldehyde in PBS to each well. Incubate cells to fix for 20 min at room temperature.
4. Wash each well with $1\times$ PBS, three times for 5 min.
5. Incubate cells in blocking buffer for 5–10 min at room temperature.
6. Dilute the primary antibody (two primary antibodies could be added at the same time, but they need to be of different origin; e.g., one rabbit and another mouse) in 100–200 µl blocking buffer, according to the recommended dilution.
7. Use a different 6-well dish to incubate the cells with the antibodies. Soak filter paper (3-cm diameter circles) in $1\times$ PBS and place them into the wells; this is necessary to maintain the humidity in the chamber. Put the coverslips on top of the soaked filter paper. Add the primary antibody carefully to cover the entire coverslip. Incubate at room temperature for 1–2 h.
8. Wash with $1\times$ PBS for 5 min, three times, each well.
9. Add the secondary antibody (fluorochrome-conjugated) in blocking buffer and incubate for 30–45 min at room temperature in the dark. For dual staining, the fluorophores need to have different emissions spectra for each individual antibody (e.g., FITC at 525 nM and phycoerythrin at 578 nM). There are a set of very sensitive and stable Alexa dyes (Molecular Probes, now part of Invitrogen Corporation, CA, USA); consult *The Handbook — A Guide to Fluorescent Probes and Labeling Technologies* for a comprehensive resource for fluorescence technology and its applications (http://probes.invitrogen.com/handbook/).
10. Wash with $1\times$ PBS for 5 min, three times, each well.
11. Pick up coverslips with a forceps and drain away excess $1\times$ PBS.
12. For mounting, add a drop of Vectashield to a clean microscope slide and gently lay the coverslip on top.
13. Remove excess Vectashield by blotting with Kimwipe and seal with nail polish.
14. After adding the secondary antibody, keep slides in the dark at all times. A similar protocol can be used for tissue sections (*see* **Note 1**).
15. Store slides in a –20°C freezer.

3.2. Caspase-3 Activity Determination: Colorimetric Assay

A simple colorimetric assay can measure the release of the chromogenic group from the synthetic substrate, most commonly pNA by activated caspases. Ac-DEVD-pNA is most frequently used, with the cleaved pNA being monitored colorimetrically through its absorbance at 405–410 nM. Although DEVD-based substrates are called caspase-3-specific, they are in fact cleaved by most caspases, with caspase-3 being the most efficient. In vitro titration experiments and/or use of specific inhibitors may be required to distinguish the activity of various caspases. Other DNA substrates are available for several other caspases.

1. Wash cells (1×10^6) with cold $1\times$ PBS and resuspend them in 50 µl cold lysis buffer, vortex, and keep on ice for 30 min.

2. Centrifuge the cell lysates at 12,000 × *g* for 10 min at 4°C, collect the supernatant in fresh tubes, and assay the protein concentration for each sample. Keep on ice.
3. To a 96-well plate, add reaction buffer, caspase substrate (100 μM final concentration), and 20–50 μl cell lysates for a final 200 μl reaction volume.
4. Incubate samples at 37°C for 1–2 h and monitor the enzyme-catalyzed release of pNA at 405 nM using a microtiter plate reader.

3.3. Immunoblot Detection of Active Caspase-3

In most cases, the 32-kDa protein procaspase (inactive caspase)-3 is activated to active caspase-3 (p17 and p12) that can be detected by western blot analysis. One of its many cellular substrates is PARP-1. In the apoptotic cells, PARP-1 (110 kDa) is cleaved to form two truncated fragments; most frequently, the 86-kDa fragment being detected by the available commercial antibodies.

3.3.1. Immunoblot Analyses

1. Collect the treated and untreated cells (1 × 10⁶) by centrifugation (500 × *g* for 5 min). Decant the medium and resuspend the cell pellet in cold 1× PBS very gently and spin it down (500 × *g* for 5 min). Decant the supernatant and repeat the process one more time. Remove 1× PBS carefully without disturbing the cell pellet.
2. Lyse the cells in a lysis buffer, with the cells incubated for 30 min on ice with occasional vortexing.
3. Centrifuge the cells for 15 min at 15,000 × *g* and collect the supernatants.
4. Perform the protein estimation of these samples using a spectrophotometric method using the Bio Rad Protein Assay reagent (working solution, 1:10 dilution) at 595 nM (1–2 μl sample will be mixed with 1 ml diluted Bio Rad Protein Assay reagent) and measure the concentration of unknown samples from the BSA standard curve (the curve can be drawn from the spectrophotometric readings of known concentrations of BSA).
5. Load 50–100 μg proteins as well as the protein standard marker on an 8–12% SDS–PAGE gel to separate the proteins under denaturing conditions.
6. Transfer the proteins to a nitrocellulose membrane by either the wet or the semi-dry transfer method.
7. Block the membrane with 5% milk for 1 h at room temperature or overnight at 4°C.
8. Incubate the membrane with primary antibodies (PARP-1 and active caspase-3) for 2 h at room temperature or overnight at 4°C (following the company's recommended dilution) (*see* **Note 2.**)
9. Wash the blot three times with 1× PBST at room temperature at 10-min intervals.
10. Add the appropriate secondary antibody (anti-mouse or anti-rabbit depending on the primary antibody) with a 1:2000 dilution to the blot and incubate for 1–1.5 h at room temperature.

11. Wash the blot five times with 1× PBST at room temperature at 10-min intervals.
12. Wash the blot with double-distilled water for a very short time to get rid of Tween 20 and develop it using chemiluminescent reagents, such as lumiglo or ECL, following the company's suggested protocol.

3.4. Fluorescein-Labeled CaspaTag Pan-Caspase In Situ Assay Kit

This is a useful approach to detect the active caspase in individual cells. The methodology is based on the FLICAs. The inhibitors are cell permeable and non-cytotoxic. This kit contains a carboxyfluorescein-labeled fluoromethyl ketone peptide inhibitor of caspase (FAM-VAD-FMK), which generates a green fluorescence. This probe covalently binds to a reactive cysteine residue on the large subunit of the active caspase heterodimer, thereby inhibiting further enzymatic activity after being taken up into cells. The green fluorescence provides the direct measure of the active caspase present in the cell, and it can be quantitated by flow cytometry, although it can be also analyzed by immunofluorescence to provide information on single cells that can be visualized by microscopy.

1. Transfer approximately 300 µl each cell suspension (~10^6 cells) to sterile tubes.
2. Add 10 µl freshly prepared 30× FLICA reagent and mix cells by flicking the tubes.
3. Incubate tubes for 1 h at 37°C under 5% CO_2, protecting tubes from light. Swirl tubes once or twice during this time to gently resuspend the settled cells.
4. Add 2 ml 1× wash buffer to each tube and mix gently.
5. Centrifuge the cells at $400 \times g$ for 5 min at room temperature.
6. Remove the solution carefully and discard the supernatant. Gently vortex the cell pellet to disrupt any cell-to-cell clumping.
7. Wash the cells with 1 ml 1× wash buffer.
8. Resuspend the cell pellet in 400 µl 1× wash buffer.
9. For bicolor analysis, add 2 µl PI solution to one cell suspension. Set aside a second suspension without PI.
10. For single-color analysis, keep the samples on ice and analyze on the FL1 channel. Otherwise, 40 µl fixative can be added and cells can be stored at 2–8°C, protected from light. However, for bicolor analysis, cells to be analyzed with PI cannot be fixed. Instead, they have to be analyzed immediately on the FL1 channel of FACS for fluorescein and FL2 channel for red fluorescence (*see* **Note 3**).

4. Notes

1. Formalin-fixed and paraffin-embedded mouse *(17)* or patient-derived human *(18)* tissue sections can be also examined. The slides are deparaffinized with xylene and graded alcohol and treated with citrate buffer (pH 6) for 20 min for antigen

retrieval before incubation with primary antibodies. Sections are counterstained with hematoxylin before being examined under the microscope. Immunohisto-chemistry for caspase-3 can be combined with the in situ detection of apoptotic cells by terminal deoxynucleotide transferase-mediated dUTP nick-end labeling (TUNEL) *(18)*. TUNEL or Comet assays are methods that detect DNA strand breaks that are associated with apoptosis. The samples are first immunostained for caspase-3 and, after washing in PBS, with a horseradish peroxidase (HRP)-linked secondary antibody. Immunoreactivity is visualized by a 10-min incubation with the HRP substrate diaminobenzidine. After staining for caspase-3, the same slides are then processed for in situ detection and localization of apoptosis at the level of single cells. Sections are then stained with anti-fluorescein antibodies linked with alkaline phosphatase, developed with Fast Red substrate and counterstained with hematoxylin.

2. The molecular weight of native PARP-1 is 110 kDa and that of cleaved PARP-1 is 86 kDa. The molecular weight of procaspase-3 is 32 kDa, whereas active caspase-3 migrates at 17 as well as 12 kDa. Some antibodies recognize only the pro-form of caspase-3, some recognize only the active form, and some can recognize both. The primary antibodies can be reused for a couple of times if they are stored at 4°C in the presence of sodium azide (0.01%, w/v).

3. Flow cytometry analysis can be done with single-color (FLICA alone) or dual-color staining (FLICA and PI). It is recommended that induced and non-induced samples be run for each labeling condition (unlabeled, FLICA-labeled, PI-labeled, and FLICA/PI-labeled).

Acknowledgments

This work was supported by a research grant from the National Institutes of Health to AA (CA81504).

References

1. Danial, N. N., and Korsmeyer, S. J. (2004) Cell death: critical control points. *Cell* **116**, 205–19.
2. Green, D. R., and Evan, G. I. (2002) A matter of life and death. *Cancer Cell* **1**, 19–30.
3. Thompson, C. B. (1995) Apoptosis in the pathogenesis and treatment of disease. *Science* **267**, 1456–62.
4. Evan, G. I., and Vousden, K. H. (2001) Proliferation, cell cycle and apoptosis in cancer. *Nature* **411**, 342–8.
5. Chen, Q., Gong, B., and Almasan, A. (2000) Distinct stages of cytochrome c release from mitochondria: evidence for a feedback amplification loop linking caspase activation to mitochondrial dysfunction in genotoxic stress induced apoptosis. *Cell Death Differ* **7**, 227–33.

6. Gong, B., and Almasan, A. (2000) Apo2 ligand/TNF-related apoptosis-inducing ligand and death receptor 5 mediate the apoptotic signaling induced by ionizing radiation in leukemic cells. *Cancer Res* **60**, 5754–60.
7. Kumar, S. (1999) Mechanisms mediating caspase activation in cell death. *Cell Death Differ* **6**, 1060–6.
8. Walker, N. P., Talanian, R. V., Brady, K. D., Dang, L. C., Bump, N. J., Ferenz, C. R., Franklin, S., Ghayur, T., Hackett, M. C., Hammill, L. D., and et al. (1994) Crystal structure of the cysteine protease interleukin-1 beta-converting enzyme: a (p20/p10)2 homodimer. *Cell* **78**, 343–52.
9. Rotonda, J., Nicholson, D. W., Fazil, K. M., Gallant, M., Gareau, Y., Labelle, M., Peterson, E. P., Rasper, D. M., Ruel, R., Vaillancourt, J. P., Thornberry, N. A., and Becker, J. W. (1996) The three-dimensional structure of apopain/CPP32, a key mediator of apoptosis. *Nat Struct Biol* **3**, 619–25.
10. Wilson, K. P., Black, J. A., Thomson, J. A., Kim, E. E., Griffith, J. P., Navia, M. A., Murcko, M. A., Chambers, S. P., Aldape, R. A., Raybuck, S. A., and et al. (1994) Structure and mechanism of interleukin-1 beta converting enzyme. *Nature* **370**, 270–5.
11. Thornberry, N. A., Rano, T. A., Peterson, E. P., Rasper, D. M., Timkey, T., Garcia-Calvo, M., Houtzager, V. M., Nordstrom, P. A., Roy, S., Vaillancourt, J. P., Chapman, K. T., and Nicholson, D. W. (1997) A combinatorial approach defines specificities of members of the caspase family and granzyme B. Functional relationships established for key mediators of apoptosis. *J Biol Chem* **272**, 17907–11.
12. Hengartner, M. O. (2000) The biochemistry of apoptosis. *Nature* **407**, 770–6.
13. Fischer, U., Janicke, R. U., and Schulze-Osthoff, K. (2003) Many cuts to ruin: a comprehensive update of caspase substrates. *Cell Death Differ* **10**, 76–100.
14. Gong, B., Chen, Q., Endlich, B., Mazumder, S., and Almasan, A. (1999) Ionizing radiation-induced, Bax-mediated cell death is dependent on activation of cysteine and serine proteases. *Cell Growth Differ* **10**, 491–502.
15. Mazumder, S., Chen, Q., Gong, B., Drazba, J. A., Buchsbaum, J. C., and Almasan, A. (2002) Proteolytic cleavage of cyclin E leads to inactivation of associated kinase activity and amplification of apoptosis in hematopoietic cells. *Mol Cell Biol* **22**, 2398–409.
16. Mazumder, S., Gong, B., and Almasan, A. (2000) Cyclin E induction by genotoxic stress leads to apoptosis of hematopoietic cells. *Oncogene* **19**, 2828–35.
17. Ray, S., and Almasan, A. (2003) Apoptosis induction in prostate cancer cells and xenografts by combined treatment with Apo2 ligand/tumor necrosis factor-related apoptosis-inducing ligand and CPT-11. *Cancer Res* **63**, 4713–23.
18. Masri, S. C., Yamani, M. H., Russell, M. A., Ratliff, N. B., Yang, J., Almasan, A., Apperson-Hansen, C., Li, J., Starling, R. C., McCarthy, P., Young, J. B., and Bond, M. (2003) Sustained apoptosis in human cardiac allografts despite histologic resolution of rejection. *Transplantation* **76**, 859–64.

3

Flow Cytometry Enumeration of Apoptotic Cancer Cells by Apoptotic Rate

David Diaz, Alfredo Prieto, Eduardo Reyes, Hugo Barcenilla, Jorge Monserrat, and Melchor Alvarez-Mon

Summary

Most authors currently quantify the frequency of apoptotic cells in a given phenotypically defined population after calculating the apoptotic index (AI), that is, the percentage of apoptotic cells displaying a specific lineage antigen (LAg) within a population of cells that remain unfragmented and retain the expression of the LAg. However, this approach has two major limitations. First, apoptotic cells fragment into apoptotic bodies that later disintegrate. Second, apoptotic cells frequently lose, partially or even completely, the cell surface expression of the LAg used for the identification of specific cell subsets. This chapter will describe a flow cytometry method to calculate the apoptotic rate (AR) that takes into account both cell fragmentation and loss of LAg expression on measurement of apoptosis using flow cytometry ratiometric cell enumeration that emerges as a more accurate method of measurement of the occurrence of apoptosis in normal and tumoral cell cultures.

Key Words: Apoptosis; apoptotic rate; apoptotic index; cell enumeration; accurate apoptosis measurement; microbeads; annexin V; antigen loss; cell fragmentation.

1. Introduction

1.1. Apoptosis Measurement

The initial methods developed for the *in vitro* quantification of apoptosis measured phenomena associated with apoptosis in cultures at the population level, such as the assessment of nucleosomal DNA fragmentation after gel

From: *Methods in Molecular Biology, vol. 414: Apoptosis and Cancer*
Edited by: G. Mor and A. B. Alvero © Humana Press Inc., Totowa, NJ

electrophoresis *(1–3)*. However, it soon became clear that individual cells undergo apoptosis in a heterogeneous and asynchronous manner *(4)*. It was therefore realized that the accurate measurement of apoptosis required methods that could identify apoptotic events at the single-cell level *(5–11)*. These methods revealed the heterogeneity of the apoptotic process to be correlated with cell phenotype—at least to a certain extent *(12)*. The ongoing development of flow cytometric techniques eventually made it possible to simultaneously identify and quantify apoptotic cells phenotypically defined by the expression of their surface lineage antigens (LAgs).

The relevance of apoptosis has promoted active research into new methods of detecting these subcellular lesions at the single-cell level in complex cell mixtures both *ex vivo* and in cultured cells *(5,7,13)*. A good example is the use of annexin V for the detection of early apoptotic cells. In such cells, PS from the inner side of the plasma membrane to the outer membrane leaflet, where it becomes exposed *(14–16)*. It can then be bound by annexin V, a phagocyte membrane protein *(17)*. The availability of fluorochrome-labeled recombinant soluble annexin V provides a useful tool for detecting and quantifying early apoptotic cells by flow cytometry *(7,8,14)*. The annexin V-labeling method can be improved by the staining with the vital dye 7-amino-actinomycin D (7AAD) *(11)* to identify early and late apoptotic cells and necrotic ones. Cell washing, and choice of resuspension buffer, can affect the accuracy of measurements of apoptosis. It has been shown that wash cycles not only cause cell loss but also affect the viability of cells as well as the precision of repeat measurements. Therefore, wash cycles should be reduced to a minimum, which also reduces the time required for sample preparation.

1.2. Discrimination Between Whole Cells and Cell Fragments by Flow Cytometry

7AAD labeling can be used to discriminate between either viable and apoptotic whole cells or cell fragments. Apoptosis led to the fragmentation of apoptotic cells into apoptotic bodies under different culture conditions (*see* **Fig. 1**). Compared with whole cells, apoptotic bodies are smaller, consistent with the notion that one cell generates several apoptotic bodies. The inclusion of the latter in the cell analysis gate, and their subsequent consideration as apoptotic cells, results in an overestimation of the frequency of apoptosis, and therefore, discrimination between cells and apoptotic bodies is critical for accurate measurement of apoptosis.

As shown in **Fig. 1**, the discrimination between either viable and apoptotic whole cells or cell fragments was achieved by the analyses of both their

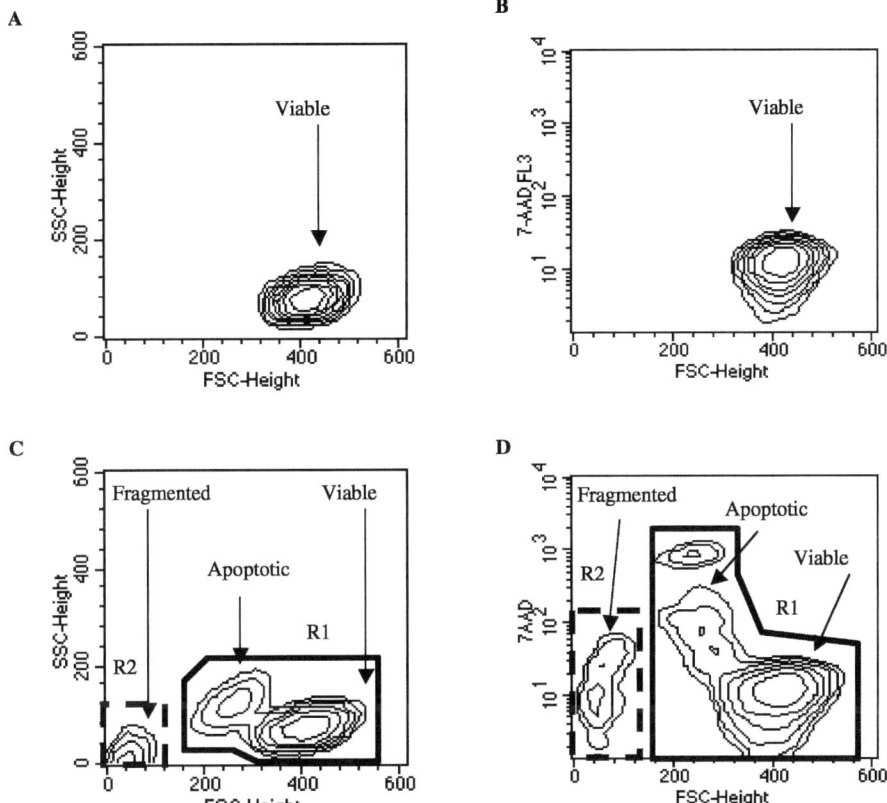

Fig. 1. Flow cytometry approach used to discriminate whole cells from apoptotic bodies by gating in 7-amino-actinomycin D (7AAD)/FSC bivariate dot-plots. Freshly purified CD19$^+$ lymphocytes were labeled with CD19-APC, annexin V-FITC, and 7AAD. Flow cytometry analysis was performed before and after 24 h of culture. The experiment was repeated six times. Panels **A** and **B** show SSC/FSC and 7AAD/FSC bivariate contour plots of freshly purified B cells. Panels **C** and **D** show how whole cells (R1) were differentiated from apoptotic bodies (R2, 7AAD$^-$, and lower FSC signal than the lower limit of the 7AAD$^+$ apoptotic cells) through combined analysis of the FSC/SSC/7AAD characteristic of the events measured.

bivariate profiles of size (FSC)/DNA staining with 7AAD (right panels) and their FSC/granularity (SSC) distribution (left panels). Contour plots in the top panels show that freshly purified CD19$^+$ lymphocytes formed a homogeneously sized population of viable cells that uniformly excluded 7AAD. After 24 h of culture (bottom panels), both apoptotic cells and apoptotic bodies emerged, but the remaining subset of viable B cells maintain the characteristics from the original fresh B lymphocytes, as they shared their FSC/SSC features (panel C) and did not take up 7AAD (contour levels under viable cell arrows in the panels C and D). In contrast, apoptotic cells showed a reduced FSC and a slightly increased SSC (panel C) that was coincident with variable 7AAD staining related to progression into late apoptosis(panel D). Finally, apoptotic bodies showed markedly smaller FCS and SSC signals than did live or apoptotic whole cells independently of their occasionally weak 7AAD staining (panels C and D). Thereafter, an event was considered to correspond to a whole cell when it provided an FSC signal greater than the lower limit of 7AAD$^+$ apoptotic cells (insert of continuous line boxes in bottom panels). Using these criteria, whole apoptotic cells were clearly distinguishable from apoptotic bodies (inserts of discontinuous line boxes in bottom panels).

1.3. Apoptosis Quantification

Most authors currently quantify the frequency of phenotypically defined apoptotic cells after calculating the apoptotic index (AI), that is, the percentage of apoptotic cells displaying a specific LAg within a population of cells that remain unfragmented and retain the expression of the LAg *(18–20)*. However, this approach has two major limitations. First, apoptotic cells fragment into apoptotic bodies that later disintegrate. This leads to an underestimation of the percentage of apoptotic cells if the debris is excluded from the gates for cell analyses or, alternatively, to the overestimation of apoptosis if several apoptotic bodies derived from a single cell are misinterpreted as individual apoptotic cells *(21)*. Second, apoptotic cells frequently lose, partially or even completely, the cell surface expression of the LAg used for the identification of specific cell subsets *(22–24)*; this means that the apoptotic cells from one phenotypically defined cell subset that loss the expression of their characteristic LAg can no longer be identified as targets in the apoptosis quantification, which leads to miscalculations *(25)*.

The limitations of current flow cytometric approaches for evaluating apoptosis warrant the development of a new multiparameter method that (i) identifies and quantifies cells suffering apoptotic lesions in earlier stages of apoptosis, (ii) discriminates live, necrotic, and apoptotic cells in a time frame within the death program that is well ahead of LAg loss and the generation of cell debris, and (iii) extends AI to provide an estimate of the number of cells that have undergone apoptosis and its relation to the number of seeded cells: the apoptotic rate (AR).

The AR overcomes the limitations of current flow cytometric techniques that do not use internal standards to determine absolute numbers. In previously described methods *(6,7,9–11)*, AI has been used to measure the proportion of apoptotic cells in relation to the total number of detectable cells in the test tube at the end point of the cell culture assay. The enumeration of apoptotic cells by the AR reflects the proportion of cells that have undergone apoptosis in relation to the total number of cells seeded at the start point of the cell culture assay. This makes the estimation of the incidence of apoptosis more valid, as current methods ignore late apoptotic cells that have suffered LAg loss or fragmentation into apoptotic bodies. Therefore, the AR is a more sensitive indicator of apoptosis than the widely used AI.

The ability to accurately and sensitively determine the number and population of cells undergoing apoptosis will allow great advances in evaluating new therapies targeted at inducing or inhibiting apoptosis. In addition, it could provide an early marker of therapeutic outcome, enabling clinicians to quickly determine if, for example, a new chemotherapeutic agent is successfully targeting neoplastic cells or if these are resistant to the therapy. The ease of use of flow cytometric techniques allows apoptosis to be used as a clinical parameter. Well-defined interpretations of results such as AR will help develop the use of apoptosis as a marker in making clinical decisions.

A limitation of the proposed method is that AR can only be properly applied in time frames in which the *in vitro* cell proliferation do not alter significantly the number of cells in the culture. This time frame depends on the rate of proliferation of the studied cells. If the cells do not proliferate (i.e., B-chronic lymphocytic leukemia cells), then apoptosis can be measured by AR at 24 h or even 48 h of culture. When cells proliferate vigorously (i.e., certain tumor cell lines) is necessary to perform the apoptosis assays after shorter periods of culture (3–6 h) to avoid interference of proliferative processes on the quantification of cell loss by apoptosis. In any case, even in conditions in which apoptosis and growth simultaneously occur, methods that enumerate apoptotic cells provide more information than those that only provide relative proportions of apoptotic cells.

In summary, apoptosis cannot be accurately quantified by simply taking into account the percentage of cells that show apoptotic lesions. Single-cell approaches must therefore be used with care if occurrence of apoptosis is to be accurately evaluated and should take into account absolute cell enumeration through the use of an internal microbead standard and the calculation of the AR.

2. Materials

2.1. Equipment

1. Sterile 50-ml conical tubes (Becton & Dickinson Biosciences, San José, CA).
2. 5-ml polystyrene round-bottom tubes 12 mm × 75 mm style (Becton & Dickinson Biosciences).
3. 96-well flat-bottom culture plates (Sero-Wel, Viví Sterilin Ltd, Stone, Staffs, UK).
4. Neubauer chamber (Brand, Wertheim, Germany).
5. FACSCalibur flow cytometer (Becton & Dickinson Biosciences).

2.2. Reagents

1. RPMI 1640 (Biowhittaker Products, Verviers, Belgium).
2. Complete medium: RPMI 1640 supplemented with 10% heat-inactivated fetal calf serum (Gibco, Grand Island, NY), 25 mM HEPES (Biowhittaker Products), and 1% penicillin–streptomycin (Biowhittaker Products).
3. 7AAD (Sigma, St. Louis, MO). Highly toxic.
4. Ca^{2+}-binding buffer. Annexin V-binding buffer containing Ca^{2+} (HEPES 10 mM, NaCl 150 nM, $MgCl_2$ 1 mM, $CaCl_2$ 1.8 mM, and KCl 5 mM; pH adjusted to 7.4; Sigma).
5. Annexin V-FITC (Bender MedSystem, Vienna, Austria).
6. 6-μm CALIBRITE microbeads (Becton & Dickinson Biosciences).
7. Gelatin (Sigma).
8. Trypan blue (0.1%, Sigma). Highly toxic.
9. Staurosporine (0.5×10^{-6} M, Sigma). Highly toxic.
10. Cycloheximide (10^{-3} M, Sigma). Highly toxic.
11. Phytohemagglutinin (2 μg/ml, Difco Lab, Detroit, MI). Highly toxic.
12. T-Cell Expander (Dynal, Oslo, Norway).

3. Methods

The methods described below outline (i) preparation of the microbeads, (ii) preparation of the cell suspension, (iii) preparation of the culture, (iv) preparation of the basal condition, (v) acquisition of cells after culture, and (vi) calculation of the AR.

3.1. Preparation of the Microbeads

One of the major problems of the use of microbeads in flow cytometric enumeration of cells is the possibility of adherence. Owing to this, we need to block the adherence of microbeads to both the tube and the own microbeads by using gelatin in the solution used to dilute the microbeads. Another factor is the sedimentation of microbeads in the tube. Just before adding the microbeads to the cell sample, we need to vortex vigorously the microbead solution.

1. In a 50-ml conical tube, prepare a volume (μl) of Ca^{2+}-binding buffer equal to 100 $\times$ number of sample tubes we will use (*see* **Note 1**).
2. Add gelatin 0.05% (w/v).
3. Heat the tube in a thermal bath to 37°C for 30 min.
4. Keep at room temperature for 30 min.
5. Add CALIBRITE microbeads to the 50-ml tube to prepare a 1/100 (v/v) dilution.
6. Vortex the 50-ml tube during 1 min.
7. Storage at 4°C in a refrigerator until use.

3.2. Preparation of the Cell Suspension

A cell suspension of tumor cells in complete medium must be obtained. This suspension could be homogeneous (i.e., tumoral cell line) or heterogeneous (i.e., peripheral blood mononuclear cells from a patient suffering leukemia or tumor cells obtained from a tumor biopsy). All the protocol must be performed using sterile material in a laminar flow chamber.

1. Take 20 μl cell suspension and dilute it with 20 μl trypan blue (*see* **Note 2**).
2. Mix gently and count the viable cells (cells without blue staining) in a Neubauer chamber.
3. Adjust the cells to a cell concentration of 0.5×10^6 viable cells/ml.

3.3. Preparation of the Culture

1. Add 100 μl complete medium into three wells (triplicate) in 96-well flat-bottom culture plates (*see* **Note 3**) and into three 5-ml polystyrene round-bottom tubes.
2. Add 100 μl diluted cells into the wells with complete medium (*see* **Note 4**) and into three 5-ml polystyrene round-bottom tubes to make the basal condition.
3. Culture the plate at 37°C in 5% CO_2 (*see* **Note 5**).

3.4. Preparation of the Basal Condition

1. Add to the tubes a combination of monoclonal antibodies labeled in FL-2 (i.e., phycoerythrin) and FL-4 (i.e., allophycocyanin) (*see* **Note 6**).
2. Incubate cells at 4°C in the dark for 20 min.
3. Centrifuge cells at 300 $\times$ *g* and 4°C for 5 min and decant the supernatant (*see* **Note 7**).
4. Resuspend cells and add 100 μl Ca^{2+}-binding buffer.
5. Add 6 μl annexin V-FITC diluted 1/5 in Ca^{2+}-binding buffer at 4°C in the dark for 10 min.
6. Add 100 μl prepared microbeads (remember to make a vigorous vortexing of the microbead solution before adding it to the cell suspension).
7. Add 100 ml 7AAD diluted in Ca^{2+}-binding buffer to a final concentration of 2.5 μg/ml and wait for 3–5 min (*see* **Note 6**).

8. Acquire the cell tubes in a four-color flow cytometer (*see* **Note 8**). Make a cell gate around microbeads and adjust the number of acquired microbeads (i.e., 2000 microbeads) to simplify the calculations to obtain the AR.

3.5. Acquisition of Cells After Culture

1. Take out the volume of each well with a micropipette (*see* **Note 9**) and add it into 5-ml polystyrene round-bottom tubes.
2. Prepare the 24 h condition like in **Subheading 3.4** (*see* **Note 10**).

3.6. Calculation of the AR

The calculation of AR consists of two sequential steps. First, the number of events corresponding to cells that have finished the apoptotic process and have undergone fragmentation into apoptotic bodies or have completely lost the expression of surface markers is calculated from the difference between the number of events corresponding to seeded cells and that of cells that remain in culture and are LAg^+ after challenge. Second, we sum to this number the number of events corresponding to annexin V^+ cells and calculate the apoptosis occurrence with respect to the total number of seed cells.

1. $NFC = NSC - NRC$,
   where *NFC*, events corresponding to fragmented cells or that completely lost the expression of their LAg; *NSC*, events corresponding to seeded cells; and *NRC*, events corresponding to remaining cells, which includes both annexin V^+ and annexin V^- cells.
2. The AR is then calculated by the following equation (*see* **Notes 11** and **12**):

$$AR = \frac{NAV^+C + NFC}{NSC},$$

   where NAV^+C, events corresponding to the number of annexin V^+ cells.

4. Notes

1. Prepare an extra 10% more volume that we will need to be sure that we will have enough volume for all the tests if any problem arises. The minimum volume we must prepare is 10 ml because less volume can not be shaken properly. Do not use microbeads prepared since 10 or more days ago.
2. If we have an excessive number of cells per count chamber field, we can dilute the cells in trypan blue until obtain the proper dilution. A cell count per field between 30 and 120 cells allows accurate counting.
3. This is to measure spontaneous apoptosis. We can induce apoptosis by a several kind of apoptogens such as etoposide, staurosporine, or cycloheximide or even study the activation-induced cell death induced by phytohemagglutinin or microbeads coated with anti-CD3 and anti-CD28 antibodies.

4. Critical step. It is very important to seed cells carefully because it affect so much to cell enumeration.

5. The time of culture should be adjusted depending on the apoptosis and cell growth properties of the tumor cells in culture. If the cells grow quickly, then the time frame to measure apoptosis should be sorter.

6. If we want to make several different labeling of the cells with different combinations of antibodies, we must prepare three culture wells and three 5-ml polystyrene round-bottom tubes for each combination to assess the precision of apoptosis measurement. If we do not want to label with 7AAD, then we can add an additional monoclonal antibody labeled in FL-3 channel (i.e., peridinin chlorophyll protein conjugate).

7. Critical step. It is very important to decant cells carefully because it affect so much to cell enumeration because of cell loss.

8. We can label the cells with more or less fluorochrome-labeled antibodies depending on the technical characteristics of our flow cytometer.

9. Critical step. It is very important to take out cells carefully because it affect so much to cell enumeration because of cell loss. We must take out all the volume of the well.

10. It is critical to use for all the tests for a given experiment the same microbead solution for the reference of the ratiometric enumeration.

11. It should be noted that we can calculate the AR of a cell subpopulation defined by the expression of a cell marker and not only the total AR.

12. It is possible to make a more immediate calculation of AR using the next equation:

$$AR = \frac{NSC - NVC}{NSC},$$

where *NSC*, events corresponding to seeded cells; and *NVC*, events corresponding to viable cells after culture (number of annexin V$^-$ cells).

References

1. Wyllie, A.H. (1980) Glucocorticoid-induced thymocyte apoptosis is associated with endogenous endonuclease activation. *Nature*, 284, 555–556.

2. Russell, J.H. and Dobos, C.B. (1980) Mechanisms of immune lysis. II. CTL-induced nuclear disintegration of the target begins within minutes of cell contact. *J. Immunol.*, 125, 1256–1261.

3. Cohen, J.J. and Duke, R.C. (1984) Glucocorticoid activation of a calcium-dependent endonuclease in thymocyte nuclei leads to cell death. *J. Immunol.*, 132, 38–42.

4. Khong, H.T. and Restifo, N.P. (2002) Natural selection of tumor variants in the generation of "tumor escape" phenotypes. *Nat. Immunol.*, 3, 999–1005.

5. Darzynkiewicz, Z., Juan, G., Li, X., Gorczyca, W., Murakami, T. and Traganos, F. (1997) Cytometry in cell necrobiology: analysis of apoptosis and accidental cell death (necrosis). *Cytometry*, 27, 1–20.

6. Herault, O., Colombat, P., Domenech, J., Degenne, M., Bremond, J.L., Sensebe, L., Bernard, M.C. and Binet, C. (1999) A rapid single-laser flow cytometric method for discrimination of early apoptotic cells in a heterogeneous cell population. *Br. J. Haematol.*, 104, 530–537.

7. van Engeland, M., Nieland, L.J.W., Ramaekers, F.C.S., Schutte, B., and Reutelingsperger, C.P.M. (1998) Annexin V-affinity assay: a review on an apoptosis detection system based on phosphatidylserine exposure. *Cytometry*, 31, 1–9.

8. Vermes, I., Haanen, C., Steffens-Nakken, H., and Reutelingsperger, C.P.M. (1995) A novel assay for apoptosis. Flow cytometric detection of phosphatidylserine expression on early apoptotic cells using fluorescein labelled annexin V. *J. Immunol. Methods*, 184, 39–51.

9. Gorczyca, W., Gong, J., and Darzynkiewicz, Z. (1993) Detection of DNA strand breaks in individual apoptotic cells by the in situ terminal deoxynucleotidyl transferase and nick translation assays. *Cancer Res.*, 52, 1945–1951.

10. Gong, J., Traganos, F., and Darzynkiewicz, Z. (1994) A selective procedure for DNA extraction from apoptotic cells applicable for gel electrophoresis and flow cytometry. *Anal. Biochem.*, 218, 314–319.

11. Schmid, I., Krall, W.J., Uittenbogaart, C.H., Braun, J., and Giorgi, J.V. (1992) Dead cell discrimination with 7-aminoactinimicin D in combination with dual color immunofluorescence in single laser flow cytometry. *Cytometry*, 13, 204–208.

12. Pantaleo, G., Graziosi, C., Demarest, J.F., Butini, L., Montroni, M., Fox, C.H., Orenstein, J.M., Kotler, D.P., and Fauci, A.S. (1993) HIV infection is active and progressive in lymphoid tissue during the clinically latent stage of disease. *Nature*, 362, 355–358.

13. Ashkenazi, A. and Dixit, V.M. (1998) Death receptors: signaling and modulation. *Science*, 281, 1305–1308.

14. Koopman, G., Reutelingsperger, C.P.M., Kuijten, G.A., Keehnen, R.M., Pals, S.T., and van Oers, M.H. (1994) Annexin V for flow cytometric detection of phosphatidylserine expression on B cells undergoing apoptosis. *Blood*, 84, 1415–1420.

15. Devaux, P.F. (1991) Static and dynamic lipid asymmetry in cell membranes. *Biochemistry*, 30, 1163–1173.

16. Zachowski, A. (1993) Phospholipids in animal eukaryotic membranes: transverse asymmetry and movement. *Biochem. J.*, 294, 1–14.

17. Fadok, V.A., Voelker, D.R., Campbell, P.A., Bratton, D.L., Cohen, J.J., Noble, P.W., Riches, D.W., and Henson, P.M. (1993) The ability to recognize phosphatidylserine on apoptotic cells is an inducible function in murine bone marrow-derived macrophages. *Chest*, 103, 102.

18. Potten, C.S. (1996) What is an apoptotic index measuring? A commentary. *Br. J. Cancer*, 74, 1743–1748.

19. Darzynkiewicz, Z. and Traganos, F. (1998) Measurement of apoptosis. *Adv. Biochem. Eng. Biotechnol.*, 62, 33–73.

20. Darzynkiewicz, Z., Bedner, E., Traganos, F., and Murakami, T. (1998) Critical aspects in the analysis of apoptosis and necrosis. *Hum. Cell*, 11, 3–12.

21. Prieto, A., Díaz, D., Barcenilla, H., García Suárez, J., Reyes, E., Monserrat, J., San Antonio, E., Melero, D., de la Hera, A., Orfao, A., and Álvarez Mon-Soto, M. (2002) Apoptotic rate: a new indicator for the quantification of the occurrence of apoptosis in cell culture. *Cytometry*, 48, 185–193.

22. Potter, A., Kim, C., Golladon, K.A., and Rabinovith, P.S. (1999) Apoptotic human lymphocytes have diminished CD4 and CD8 receptor expression. *Cell. Immunol.*, 193, 36–47.

23. Philippé, J., Louagie, H., Thierens, H., Vral, A., Cornelissen, M., and De Ridder, L. (1997) Quantification of apoptosis in lymphocyte subsets and effect of apoptosis on apparent expression of membrane antigens. *Cytometry*, 29, 242–249.

24. Diaz, D., Prieto, A., Barcenilla, H., Monserrat, J., Prieto, P., Sanchez, M.A., Reyes, E., Hernandez-Fuentes, M.P., de la Hera, A., Orfao, A., and Alvarez-Mon, M. (2004) Loss of lineage antigens is a common feature of apoptotic lymphocytes. *J. Leukoc. Biol.*, 76, 609–615.

25. Prieto, A., Reyes, E., Diaz, D., Hernandez-Fuentes, M.P., Monserrat, J., Perucha, E., Munoz, L., Vangioni, R., de la Hera, A., Orfao, A., and Alvarez-Mon, M. (2000) A new method for the simultaneous analysis of growth and death of immunophenotypically defined cells in culture. *Cytometry*, 39, 56–66.

4

Detection of Cancer-Related Proteins in Fresh-Frozen Ovarian Cancer Samples Using Laser Capture Microdissection

Dan-Arin Silasi, Ayesha B. Alvero, Jechiel Mor, Rui Chen, Han-Hsuan Fu, Michele K. Montagna, and Gil Mor

Summary

Tumors are heterogeneous structures that contain different cell populations. Laser capture microdissection (LCM) can be used to obtain pure cancer cells from fresh-frozen cancer tissue and the surrounded environment, thus providing an accurate snapshot of the tumor and its microenvironment in vivo. We describe a new approach to isolate pure cancer cell population and evaluate protein expression. The process includes immunocytochemistry, laser microdissection, and western blot analysis. Using this technique, we can detect proteins such as X-linked inhibitor of apoptosis protein (XIAP) and Fas ligand (FasL) with as little as 1000 cells.

Key Words: Laser capture microdissection; ovarian cancer; MyD88; XIAP; FasL.

1. Introduction

Proteomics has emerged as the most important tool to study biological processes in both physiological and pathological circumstances *(1)*. However, the task of studying the proteome has its share of challenges. Proteomic analysis techniques are most accurate when applied to homogenous cell populations, and growing cells in culture is the most commonly used method to procure a pure population of cells. Cancer cells in vivo, however, are incorporated in the complex tissue architecture of the tumor. As a result, they will change their protein expression in different environments or even during different life-cycle stages *(2)*.

From: *Methods in Molecular Biology, vol. 414: Apoptosis and Cancer*
Edited by: G. Mor and A. B. Alvero © Humana Press Inc., Totowa, NJ

Several methods have been developed in an attempt to characterize and measure the fluctuations of protein expression in normal or diseased heterogeneous tissues.

In 1996, Michael Emmert-Buck and colleagues at the National Institutes of Health in Bethesda, MD, developed the laser capture microdissection (LCM) system as a method to isolate specific types of cells from complex multi-cellular structures, such as tissues *(3)*.

Basically, the LCM consists of a microscope attached to a low-power laser: infrared (cold) laser for the system developed by PixCell–Arcturus or ultraviolet (cutting) laser for the Leica and P.A.L.M. systems. Tissue sections are mounted on glass slides. The slides are either precovered with a transparent polyethylene film (Leica) or an ethylene-vinyl acetate film is placed over the dry sections (Arcturus). A cell or cluster of cells is selected under the microscope, and then a focused pulse laser delivers thermal energy that melts the plastic film. When applying LCM to fresh-frozen or paraffin-embedded specimens, the original tissue morphology is preserved and contamination from surrounding unwanted cells is avoided. The process does not influence the quantity or quality of proteins contained in the specimen. As a matter of fact, LCM can be used to isolate single living cells from culture. The film with the procured cell(s) is collected into buffer for processing. Proteins expressed by these cells can be identified using amplification techniques or gel-based methods *(4)*.

One limitation of the LCM system is that it cannot procure the same amount of cells that can be obtained from cell culture. However, it is a much faster method and provides an accurate live snapshot of the cells and their microenvironment. The precision of the LCM is limited in unstained or hematoxylin and eosin (H&E)-stained tissues because of the difficulty in identifying different cell types and structures by morphology alone. This obstacle can be overcome by utilizing immunostaining protocols that allow clear identification of the desired cell types.

Most tumors analyzed in our studies were heavily infiltrated with immune cells. We have successfully utilized and recommend CD45 (panleukocyte marker) immunohistochemistry (IHC) staining to identify and then isolate immune cells from cancer cells. As an alternative, we have also used IHC staining for the CK-7 antigen, but this is appropriate only for epithelial cancers.

The focus of our study is ovarian cancer, which is the fourth leading cause of cancer-related deaths in women in the United States and the leading cause of gynecologic cancer deaths. The high mortality rate is related to the inability to detect early disease, when treatments are most effective. As a result, approximately 80% of patients are diagnosed with advanced stage disease *(5,6)*.

The method outlined in this chapter describes the detection of three proteins, X-linked inhibitor of apoptosis protein (XIAP) *(7)*, Fas ligand (FasL) *(8,9)*, and human myeloid differentiation factor 88 (hMyD88) *(10–12)* in epithelial ovarian cancer cells. These proteins represent the spectrum of protein expression from highly expressed (XIAP and FasL) to limited expression (hMyD88). This method, however, has many applications and can be used for the detection of other proteins as well.

The methods outlined below are utilized for

1. Collection, freezing, and storage of tissue specimens,
2. Preparation of the sample to be analyzed,
3. Preparation of the slides,
4. IHC staining,
5. Obtaining pure cancer cells by LCM,
6. Lysis of the dissected cells, and
7. Western blot analysis of the specimen.

2. Materials

2.1. Equipment

1. LCM system (Leica Microsystems Wetzlar GmbH, Wetzlar, Germany) equipped with VSL-337ND-S N_2 laser.
2. Cryomolds, intermediate size recommended (Miles, Elkhart, IN).
3. Microtome for cryosections.
4. Leica PEN-covered slides (Microdissect GmbH, Mittenaar, Germany) (*see* **Fig. 1A**).
5. Liquid blocker PAP pen (Daido Sangyo, Tokyo, Japan).
6. 0.5-ml PCR tubes.
7. Kodak Image Station 2000R (Eastman Kodak Company, Rochester, NY) and Kodak Molecular Imaging Software (*see* **Note 1**).
8. Non-sterile forceps and scalpel.
9. Staining dishes.

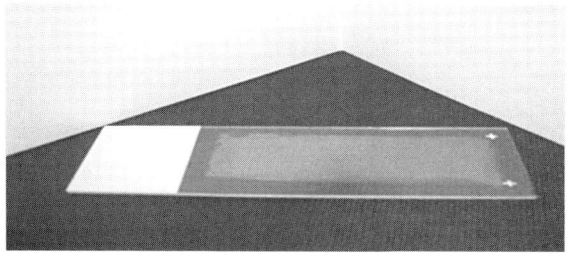

Fig. 1. Slide for sample collection and microdissection.

2.2. Reagents

2.2.1. Reagents for Tissue Preparation

1. Tissue-Tek Optimum Cutting Temperature (OCT) Compound (Sakura Finetek USA, Torrance, CA).

2.2.2. Antibodies and Reagents for Immunocytochemistry

1. Double-distilled water (ddH$_2$O).
2. Bovine serum albumin (BSA; Sigma Aldrich, St. Louis, MO).
3. Mouse anti-human leukocyte common antigen CD45 (Dakocytomation, Carpinteria, CA).
4. Mouse anti-human cytokeratin-7 (Dakocytomation).
5. Biotinylated anti-mouse IgG (H+L) secondary antibody made in horse (Vector Laboratories, Burlingame, CA).
6. Streptavidin–horseradish peroxidase (HRP) conjugate (Zymed, San Francisco, CA).
7. Diaminobenzidine tetra hydrochloride (DAB).
8. Tris [hydroxymethyl] aminomethane (American Bioanalytical, Natick, MA).
9. Mayer's hematoxylin (Sigma Diagnostics, St. Louis, MO).
10. Hydrogen peroxide (H$_2$O$_2$).
11. Ammonium hydroxide (NH$_4$OH).
12. Histosolve (Shandon, Pittsburgh, PA).
13. Ethanol (ETOH) 95%.
14 Triton X-100 (Sigma Aldrich).

2.2.3. Reagents for Cell Lysis

1. Lysis buffer [2.5% SDS, 10% glycerol, 5% β-mercapto-ethanol, 0.15 M Tris (pH = 6.8), and 0.01% bromophenol blue].

3. Methods

3.1. Collection and Freezing of the Specimen

1. Transport tumor tissue from the operating room in normal saline.
2. Examine tumor tissue obtained at the time of surgery and trim macroscopic necrosis areas or non-cancerous looking tissue.
3. Avoid unnecessary manipulation or crushing of the specimen, as this will impede the IHC staining (see **Note 2**).
4. Place the tissue in labeled cryovials, again, without excessive manipulation.
5. Immerse the cryovials in liquid N$_2$.
6. Store at −80°C.

3.2. Sample Preparation

1. Remove the cryovials from the freezer and hold it in the palm for approximately 20 s, until the tissue can be retrieved from inside the vial.
2. Avoid excessive probing of the specimen while attempting to extract it from the cryovial.
3. Cut a 3–4 mm thick section with the scalpel and position it at the center of the cryomold (*see* **Fig. 2**).
4. Pour OCT over the specimen and freeze.

3.3. Slides Preparation

See **Note 3** for details before we begin:

1. Leave the specimens in the cryostat for approximately 30 min, so they reach the cryostat chamber's temperature (we recommend –16 to –20°C).
2. Cut 8-μm sections without using the anti-roll bar and collect them on the slide using a soft brush.
3. Once a slide is full with tissue sections, fix the specimen in 95% ETOH for 30 s (*see* **Note 4**).
4. After fixing in ETOH, let the slides air-dry. If they will not be used immediately, place in airtight container and freeze without washing the OCT. When ready to use, just let them get to room temperature, wash the OCT with ddH$_2$O, and follow the protocol as for fresh sections.

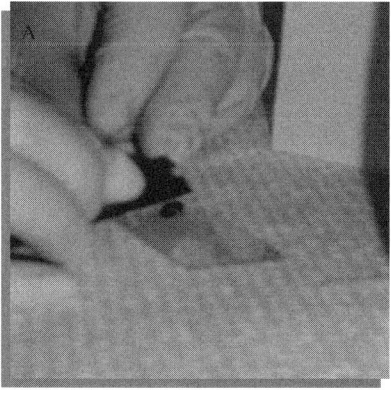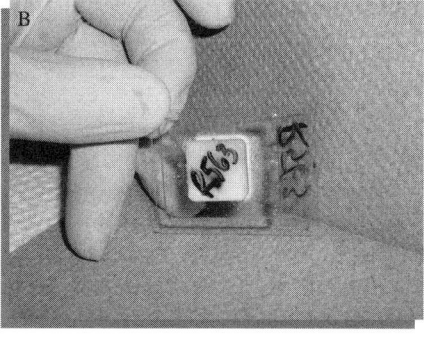

Fig. 2. Tissue preparation for sections. (**A**) Cutting a small sample. (**B**) Preparation in optimum cutting temperature (OCT).

5. If more than one slide will be prepared, fix each one and wash in ddH$_2$O to remove the OCT as soon as the tissue is cut. Then they can be placed in the moisture chamber to wait while the other slides are getting ready.

3.4. Immunostaining Protocol for Fresh-Frozen Sections in Preparation for LCM

3.4.1. Before Starting

For steps that use reagents other than ETOH, ddH$_2$O, or Histosolve, no staining dishes are needed. All reagents can be applied with a pipette, thus saving materials and time.

Antibody incubation should be accomplished in a moisture chamber.

1. Fix specimens in ETOH 95% for 30 s.
2. Wash slides in ddH2O to remove the OCT from the slides.
3. If not prepared before microtome cutting, ring area around sections with a PAP pen to preserve the amount of antibodies and reagents used.
4. Quench for endogenous peroxidase activity with 0.1% H$_2$O$_2$ in 0.1 M PB for 10 min.
5. Wash 3× with wash buffer (0.1 M PB + 0.01% Triton X-100).
6. Add 3% BSA (made in wash buffer) for 20 min to block for non-specific background.
7. Pour off the 3% BSA and add the primary antibodies (dilution 1:150 in 1% BSA/wash buffer).

    a. For CK-7, incubate for 2 h.
    b. For CD-45, incubate for 1 h.
    c. For CD-14, incubate for 1 h.

8. Wash sections with wash buffer 3× (2–3 min each).
9. Incubate with biotin-labeled secondary antibodies made in 1% BSA/wash buffer in a 1:200 dilution for 30 min.
10. Wash 1× with wash buffer, then 2× with 0.1 M Tris, pH 7.5.
11. Incubate with streptavidin–HRP for 30 min in a 1:300 dilution made in 0.1 M Tris, pH 7.5.
12. Wash 3× with 0.1 M Tris, pH 7.5.
13. Develop color with DAB in 0.1 M Tris, pH 7.5 (100 μl DAB in 1 ml Tris); add 0.03% H$_2$O$_2$ just before use (25 μl 0.03% H$_2$O$_2$ to 5 ml DAB/Tris).
14. Wash 2× with ddH$_2$O.
15. Counterstain with hematoxylin; blue with NH$_4$OH water.
16. Dehydrate in 95% ETOH for 15 min, let air-dry (2–3 min), and immerse in Histosolve for 15 min.
17. Air-dry for 15 min.
18. Wrap in plastic foil and store at –20°C.

3.5. LCM—Obtaining Pure Cancer Cells

3.5.1. Before Starting

If the slides were stored wrapped in foil at −20°C, allow them to get to room temperature before unwrapping. The following steps are described for those using a Leica Laser Microdissector.

1. Turn on the LCM system and only then launch the software by clicking "Leica Laser Microdissection" in the start menu or desktop. Switch on the Laser. It is not recommended to leave the Laser on for extended periods of time if the system is idling.
2. Place the slides inverted into the removable specimen holder (the side containing the tissue is facing down: non-contact method).
3. On the computer screen, from the "Options" tab, select "Settings" and then "Object counting." Enter the desired diameter in the provided field. This will calculate the number of objects within the outlined area. Recommended diameter: 25 μm. For ease of calculating the total number of cells selected, the box that displays the number of cells can be exported to Excel.
4. Select the drawing mode. The "Select" key needs to be pressed. The options are "Single shape" and "Multiple shapes." The latter allows us to work faster. The "Multiple shapes" mode also allows us to allocate different caps/wells to different areas, if so desired. However, do not select more than five or six shapes outside the viewable screen or else the selected areas will shift.
5. Check the "Close line(s)" option in the "Draw shape" box on the screen with the "Select" key pressed. Otherwise, "Object counting" will not work.
6. If the Laser beam is out of focus, initiate the autocalibrating sequence and follow the instructions on the screen (use the "Laser" tab, then "Calibrate").
7. For best results when cutting, we recommend the following settings (use the "Laser" tab, then "Control"):

    a. Aperture 4–6.
    b. Intensity highest 46.
    c. Speed 4–5.

8. Set the "Cut shape(s)" to standard mode.
9. Set the microscope objective to ×20 or ×40 magnification.
10. Load the PCR tubes into the LCM holder (*see* **Fig. 3**).
11. Load the tube cap(s) with 30 μl sample buffer [2.5% SDS, 10% glycerol, 5% β-mercapto-ethanol, 0.15 M Tris (pH = 6.8) and 0.01% bromophenol blue]. Load slowly, without creating too many bubbles. When cutting is completed, add another 20 μl. It is important to work fast, as the buffer evaporates almost completely after 2–3 h in an open cap. This is why we recommend working with maximum two collection caps at a time.

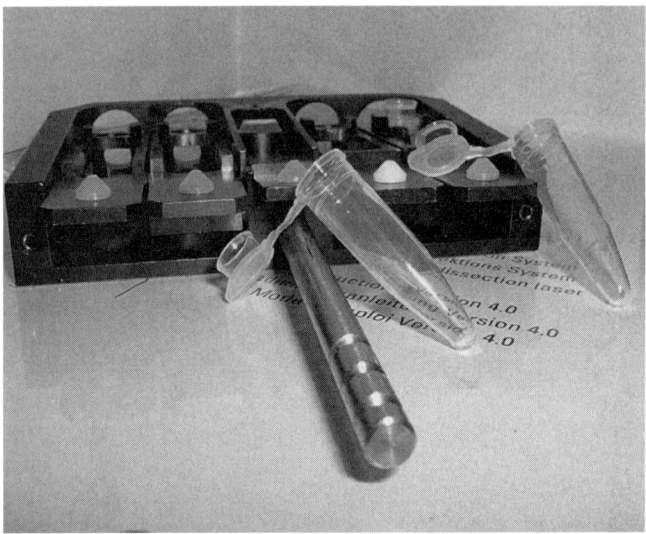

Fig. 3. Holders and tubes for sample collection.

12. Use the "Draw" option with the "Line" button activated. Trace the area that we want to dissect with the mouse cursor. Areas that are outside the viewable screen can be brought into view and then selected after moving the slide with the Smartmove controls. However, do not move the XY axis Smartmove controls during drawing of an area.

3.6. Documentation

We recommend that we document our work and save representative images (*see* **Figs 4** and **5**). This is easily accomplished by using the "Save" button.

3.7. Cell Lysis

1. Once the desired number of cells is dissected (we recommend 6000 cells for MyD88 and 1000 cells for XIAP or FasL) and another 20 μl SDS sample buffer are added to the cap, remove the PCR tube from the holder and close the cap.
2. Spin the tubes for few seconds and let them stand for 2–3 h. The SDS buffer will prevent degradation of proteins.
3. After a short centrifugation, submit the samples to five cycles of freezing (liquid N_2) and thawing (95°C), followed by boiling for 5 min.
4. Cool on ice and centrifuge again for 2 min.
5. Samples can be stored at –80°C until further use. Protein of interest can be visualized by western blot (*see* **Fig. 6**).

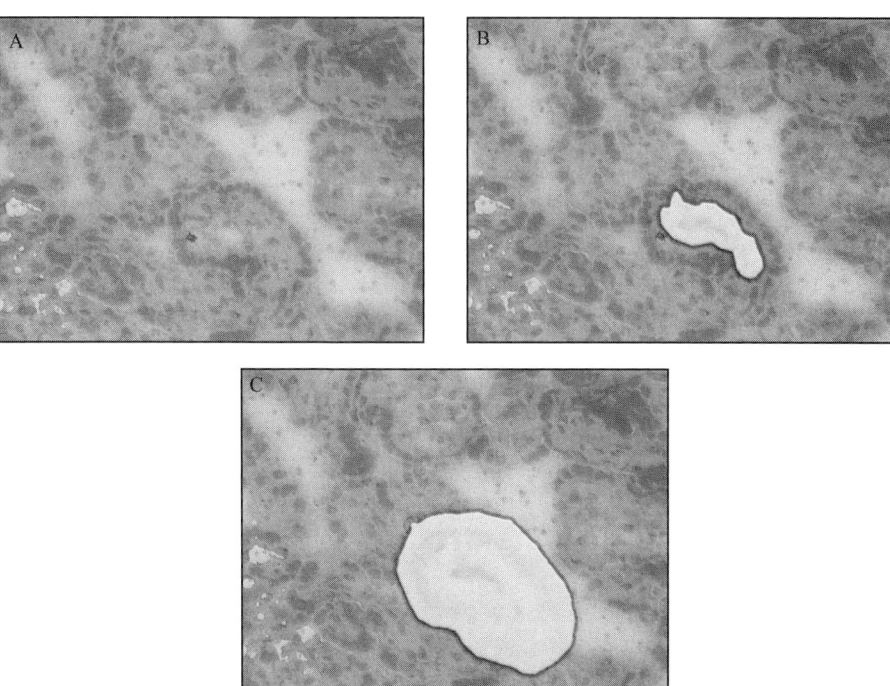

Fig. 4. Placenta samples for microdissection. (**A**) Prior cutting. (**B**) Cutting of the stroma. (**C**) Cutting of the cytotrophoblast.

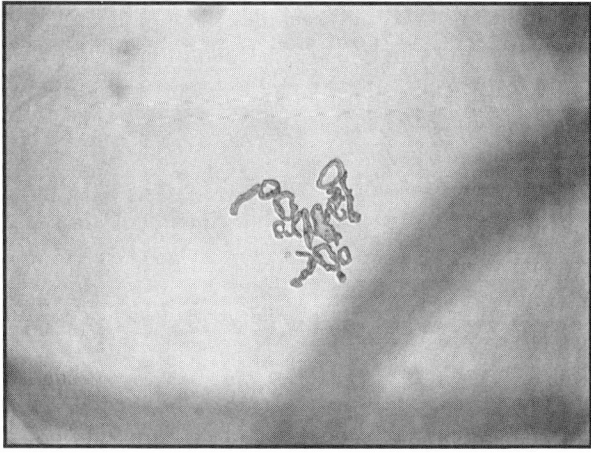

Fig. 5. Sections microdissected and collected in Ependorf tubes.

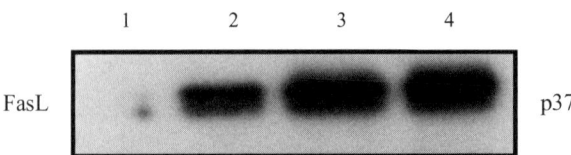

Fig. 6. Western blot analysis of microdissected cells. 1, 500 cells; 2, 1000 cells; 3, 3000 cells; 4, 5000 cells.

Highly expressed proteins such as XIAP and FasL can be detected in as little as 1000 cells, whereas MyD88 requires samples of 5000 cells.

4. Notes

1. This equipment is required only if digital imaging is preferred over standard photographic film.
2. The tissues should be frozen no longer than 20–25 min following removal from the patient.
3. Not all frozen samples are suitable for LCM. They may contain too much necrosis or too little cancer cells. To avoid a lengthy preparation of unusable samples, we recommend an H&E stain on regular glass slides before fixing the tissue sections on PEN-covered slides. A good section contains large sheets of tumor cells. This simple step will save our hours spent with the microdissector. Also, this is the last time the PEN-covered slides will be dry, so it is much easier now to score the margins with a PAP pen.
4. Fix every specimen as soon as the slide is full. Avoid spending more than a few minutes to fill up a slide. If the cut specimens are unsatisfactory, the PEN-covered slides can be recovered by washing them very gently under running water with a gloved finger. Rinse and dry. Do not blot, as this will puncture the very thin film covering the slides. Slides cannot be recovered once staining is completed.

References

1. Jones MB, Krutzsch H, Shu H, Zhao Y, Liotta LA, Kohn EC, Petricoin EF 3rd. Proteomic analysis and identification of new biomarkers and therapeutic targets for invasive ovarian cancer. *Proteomics* 2(1):76–84, 2002.
2. Agiostratidou G, Sgouros I, Galani E, Voulgari A, Chondrogianni N, Samantas E, Dimopoulos MA, Skarlos D, Gonos ES. Correlation of in vitro cytotoxicity and clinical response to chemotherapy in ovarian and breast cancer patients. *Anticancer Research* 21(1A):455–9, 2001.
3. Emmert-Buck MR, Bonner RF, Smith PD, Chuaqui RF, Zhuang Z, Goldstein SR, Weiss RA, Liotta LA. Laser capture microdissection. *Science* 274(5289): 998–1001, 1996.

4. Wang VW, Bell DA, Berkowitz RS, Mok SC. Whole genome amplification and high-throughput allelotyping identified five distinct deletion regions on chromosomes 5 and 6 in microdissected early-stage ovarian tumors. *Cancer Research* 61(10):4169–74, 2001.

5. Jemal A, Murray T, Samuels A, Ghafoor A, Ward E, Thun MJ. Cancer statistics, 2003. *CA Cancer J Clin* 2003;53:5–26.

6. Hoskins WJ, Perez CA, Young RC, Barakat RR, Markman M, Randall ME. *Principles and Practice of Gynecologic Oncology*. 4th ed. 2005.

7. Sapi E, Alvero AB, Chen W, O'Malley D, Hao XY, Dwipoyono B, Garg M, Kamsteeg M, Rutherford T, Mor G. Resistance of ovarian carcinoma cells to docetaxel is XIAP dependent and reversible by phenoxodiol. *Oncology Research* 14(11–12):567–78, 2004.

8. Abrahams VM, Straszewski SL, Kamsteeg M, Hanczaruk B, Schwartz PE, Rutherford TJ, Mor G. Epithelial ovarian cancer cells secrete functional Fas ligand. *Cancer Research* 63(17):5573–81, 2003.

9. Abrahams VM, Kamsteeg M, Mor G. The Fas/Fas ligand system and cancer: immune privilege and apoptosis. *Molecular Biotechnology* 25(1):19–30, 2003.

10. Hardiman G, Jenkins NA, Copeland NG, Gilbert DJ, Garcia DK, Naylor SL, Kastelein RA, Bazan JF. Genetic structure and chromosomal mapping of MyD88. *Genomics* 45(2):332–9, 1997.

11. Akazawa T, Masuda H, Saeki Y, Matsumoto M, Takeda K, Tsujimura K, Kuzushima K, Takahashi T, Azuma I, Akira S, Toyoshima K, Seya T. Adjuvant-mediated tumor regression and tumor-specific cytotoxic response are impaired in MyD88-deficient mice. *Cancer Research* 64(2):757–64, 2004.

12. Zeisel MB, Druet VA, Sibilia J, Klein JP, Quesniaux V, Wachsmann D. Cross talk between MyD88 and focal adhesion kinase pathways. *Journal of Immunology* 174(11):7393–7, 2005.

5

Flow Cytometric Detection of Activated Caspase-3

Richard Fox and Martine Aubert

Summary

Apoptosis (programmed cell death) is an active process that plays a critical role in multiple biologic processes from embryologic development, to lymphocyte development and selection, and homeostasis. The two major mechanisms of cell death are referred to as the intrinsic and extrinsic pathways. These pathways lead to a cascade of events that ultimately converge to the activation of an effector enzyme, caspase-3. Caspase-3 is a cysteine protease with aspartic specificity and a well-characterized effector of apoptosis or programmed cell death signaling. The pro-form of caspase-3 (p32 caspase-3) is sequestered as a zymogen, where upon proteolysis at a conserved DEVD sequence, is converted to the active (p17 caspase-3) enzyme capable of disassembling the cell. Cell death can become disregulated under various conditions and multiple disease states (e.g., viral infection, carcinogenesis, and metastasis). Sensitive and reproducible detection of active caspase-3 is critical to advance the understanding of cellular functions and multiple pathologies of various etiologies. Here, we provide two simple and reproducible methods to measure active caspase-3 in multiple cell types and conditions using a flow cytometric-based analysis.

Key Words: Apoptosis; caspase-3; immunofluorescence staining, flow cytometry.

1. Introduction

Apoptosis or programmed cell death can occur in a cell for a multitude of reasons from embryologic development to host immune responses. Apoptosis is an evolutionary conserved and genetically programmed process in multicellular organism as observed in *Caenorhabditis elegans* that results in the systematic disassembly and recycling of a cell *(1,2)*. A tremendous amount

From: *Methods in Molecular Biology, vol. 414: Apoptosis and Cancer*
Edited by: G. Mor and A. B. Alvero © Humana Press Inc., Totowa, NJ

of research has dissected much of the signaling that results in apoptosis *(3)*. Several measurable changes are known to occur in dying cells. Notably, the cell begins to shrink because of loss of cytoskeletal and structural proteins *(4–6)*, the nucleus condenses as a result of lamin cleavage *(7)*, DNA fragmentation occurs at histone sites *(3,8,9)*, phosphatidylserine translocates to the outside of the cell because of ATP depletion *(10)*, and so on. All those morphological and biochemical characteristics defining cell death are the result of a well-orchestrated cascade of events.

There are two basic mechanisms for the induction of apoptosis, the extrinsic pathway (death receptor mediated) and the intrinsic pathway (regulated at the level of the mitochondria) *(see* **Fig. 1**) *(2,11)*. The extrinsic pathway of

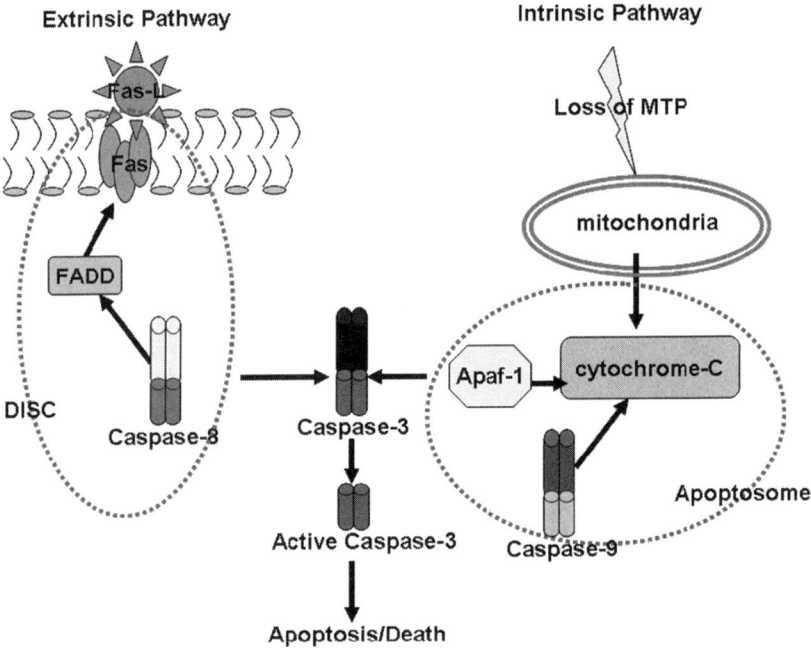

Fig. 1. Schematic of convergent extrinsic and intrinsic apoptosis signaling to caspase-3. The extrinsic pathway of apoptosis signals through cell surface molecules like Fas/FasL resulting in the recruitment of the Fas-associated death domain (FADD). FADD associates with Fas and recruits pro-caspase-8. This active complex is referred to as the death-inducing signaling complex (DISC). Caspase-3 becomes active, and cell death occurs. The intrinsic pathway occurs following loss of mitochondrial transmembrane potential (MTP) resulting in the release of cytochrome-c into the cytoplasm. Cytochrome-c complexes with the cytoplasmic protein Apaf-1 and pro-caspase-9 forming the apoptosome that results in activation of caspase-3.

apoptosis is death receptor mediated through cell surface molecules such as Fas and TRAIL (*see* **Table 1**). Ligation of a given surface molecule by a cognate ligand results in the recruitment to the receptor of molecules such as the Fas-associated death domain (FADD) leading to the subsequent recruitment of pro-caspase-8 (an initiator cysteine protease) that becomes proteolized and activated. This active complex is referred to as the death-inducing signaling complex (DISC) *(12)*. Active caspase-8 can then proteolytically process additional caspases such as caspase-3 and amplify the signaling resulting in the systematic disassembly and recycling of the cell (*see* **Fig. 1**, extrinsic stimuli).

The intrinsic pathway is regulated at the level of the mitochondria. The mitochondria serves as a sensor as well as an executioner of cell death *(11,13)*. Loss of mitochondrial transmembrane potential (MTP) results in the release of cytochrome-c into the cytoplasm. Cytochrome-c complexes with the cytoplasmic protein Apaf-1 and pro-caspase-9 forming the apoptosome. Activation of the apoptosome eventually results in the downstream activation

Table 1
Apoptosis Inducers

Apoptosis inducer	Nature and/or function	Apoptotic pathway	References
Anti-CD95	Recognizes CD95 (Fas and Apo-1), a transmembrane protein	Death receptor pathway	20
Tumor necrosis factor-α (TNF-α)	Pleiotropic inflammatory cytokine	Death receptor pathway	21
TRAIL (Apo2L)	TNF-related apoptosis-inducing ligand	Death receptor pathway	22
Etoposide (VP-16)	Topoisomerase II inhibitor	Mitochondrial pathway	23
Valinomycin	Potassium ionophore	Disrupt mitochondrial potential	24
Dexamethasone	Anti-inflammatory glucocorticoid	Mitochondrial pathway	25
Staurosporine	Inhibitor of a variety of kinases	Mitochondrial pathway	26
Doxorubicin	Anthracycline antibiotic	Mitochondrial pathway	27

of the effector caspase, caspase-3, and subsequent disassembly of the cell (*see* **Fig. 1**, intrinsic stimuli).

Both of these pathways result in the activation of caspase-3: a critical and often defining event in the process of apoptosis. Caspase-3 is a cysteine protease with aspartic specificity and is a well-characterized effector of programmed cell death signaling. Activation of caspase-3 results in the systematic disassembly of the cell through proteolytic cleavage of a multitude of substrates such as Rho-kinase (ROCK-1), an enzyme required for apoptotic body formation or the blebbing phenomenon observed in apoptotic cells *(14,16)*. The central role of caspase-3 makes it a likely target to determine whether cells are undergoing apoptosis upon stimulation with pro-apoptotic agents. In this chapter, two staining protocols are described for the detection of activated caspase-3 by flow cytometry (*see* **Note 1**.

2. Materials

2.1. Equipment

1. 15-ml polystyrene conical tubes (BD Biosciences, San Jose, CA, USA).
2. 5-ml polystyrene round-bottom tubes (BD Biosciences, San Jose, CA, USA).
3. Flow cytometer equipped with a 488-nm argon laser and a 530-nm bandpass (BP) filter for optimal detection of FITC in FL1 or a 585-nm BP filter for optimal detection of PE in FL2. For allophycocyanin (APC), a flow cytometer with a He–Ne laser at 633 nm and a filter with a 660-nm BP are recommended for optimum detection of this fluorochrome.

2.2. Reagents

1. Phosphate-buffered saline (PBS): Dulbecco's PBS without $CaCl_2$ and without $MgCl_2$.
2. 2% paraformaldehyde solution stored at 4°C ≤1 month protected from light. 2 g paraformaldehyde (electron microscopy grade, Polysciences, Warrington, PA, USA) in 100 ml PBS. Adjust the pH to 7.2 with 0.1 M NaOH or 0.1 M HCl. To dissolve paraformaldehyde, heat the solution to 70°C (without heating above 70°C) in a fume hood for approximately 1 h (*see* **Note 2**). Cool to room temperature before adjusting the pH.
3. 0.2% Tween solution is made by diluting Tween 20 (Sigma, St Louis, MO, USA) in PBS, store at 4°C ≤1 month.
4. 0.1% Tween solution obtained by diluting the 0.2% Tween solution in PBS.
5. Anti-active caspase-3 antibodies are obtained from BD Pharmingen San Jose, CA, USA: rabbit anti-active caspase-3 polyclonal antibody (cat. no. 557035), FITC-conjugated rabbit anti-active caspase-3 monoclonal antibody (cat. no. 559341), and PE-conjugated affinity purified polyclonal rabbit anti-caspase-3 (cat. no. 557091).

6. APC-conjugated F(ab′)$_2$ fragment donkey anti-rabbit IgG (H+L) (cat. no. 711-136-152) is obtained from Jackson ImmunoResearch Laboratories, West Grove, PA, USA.
7. Normal donkey serum (cat. no. 017-000-001) is obtained from Jackson ImmunoResearch Laboratories.
8. Staining solution: 0.1% NaN$_3$, 1% BSA in Dulbecco's Hanks' balanced salt solution (HBSS) without CaCl$_2$, without MgCl$_2$, and without MgSO$_4$, store at 4°C ≤1 month.

3. Methods

3.1. Induction of Apoptosis

In **Table 1** are a few examples of reagents capable of inducing apoptosis. To test whether a cell type is capable of undergoing apoptosis, the cells can be treated with various reagents known to induce apoptosis. The choice of the reagents should be determined by the nature of the cells (e.g., hematopoietic and epithelial), their origin (normal tissue or tumor), and the ability of the different reagents to induce apoptosis. In addition, to identify apoptotic pathways that may be impaired, the choice of the apoptosis inducer will be essential. Some reagents such as anti-Fas antibodies will trigger the death receptor/extrinsic pathway, whereas other reagents such as staurosporine will activate the mitochondrial/intrinsic pathway.

3.2. Detection of Activated Caspase-3 Indirect Staining

1. Collect 5×10^5 to 1×10^6 cells (*see* **Note 3**) into a 15-ml conical tube (*see* **Note 4**).
2. Resuspend the cell pellet in 1 ml 2% paraformaldehyde solution.
3. Incubate for 20–30 min on ice. The cells can be stored after this step overnight at 4°C if needed (*see* **Note 5**).
4. Centrifuge 300 g for 5 min at room temperature; discard the solution.
5. Resuspend the cell pellet in 0.5 ml 0.2% Tween solution.
6. Incubate for 15 min at 37°C to permeabilize the cells.
7. Centrifuge 300 g for 5 min at room temperature; discard the solution.
8. Wash with 1 ml PBS.
9. Centrifuge 300 g for 5 min at room temperature.
10. Resuspend the cell pellet with 50 μl staining solution and add 0.5 μl (0.25 μg) rabbit anti-active caspase-3 polyclonal antibody (for the minimum amount of antibody needed, *see* **Note 6**).
11. Incubate for 30 min at room temperature; cover in the dark.
12. Add 1 ml PBS and pellet the cells at 300 g for 5 min at room temperature.
13. Resuspend the cell pellet in 50 μl staining solution and add normal donkey serum at 1:50 dilution (same dilution as the secondary antibody, *see* **Note 7**).
14. Incubate for 10 min at room temperature.

15. Add 1 ml PBS and pellet the cells at 300 g for 5 min at room temperature.
16. Resuspend the cell pellet with 50 μl staining solution and add donkey APC-conjugated anti-rabbit antibody.
17. Incubate for 30 min at room temperature in the dark.
18. Add 1 ml 0.1% Tween solution.
19. Centrifuge 300 g for 5 min at room temperature; discard the solution.
20. Add 200–400 μl staining solution and transfer to a 5-ml round-bottom tube.
21. Analyze the cells using a flow cytometer with a He–Ne laser at 633 nm and a filter with a 660-nm BP for optimum detection of APC fluorochrome (*see* **Note 8** and **Fig. 2** for an example of expected results).

3.3. Detection of Activated Caspase-3 by Direct Staining

1. Proceed as described in **Subheading 3.2** from steps 1 to 9.
2. Resuspend the cell pellet in 50 μl staining solution + 10 μl FITC-conjugated anti-active caspase-3 or 10 μl PE-conjugated anti-active caspase-3 (*see* **Note 9**).

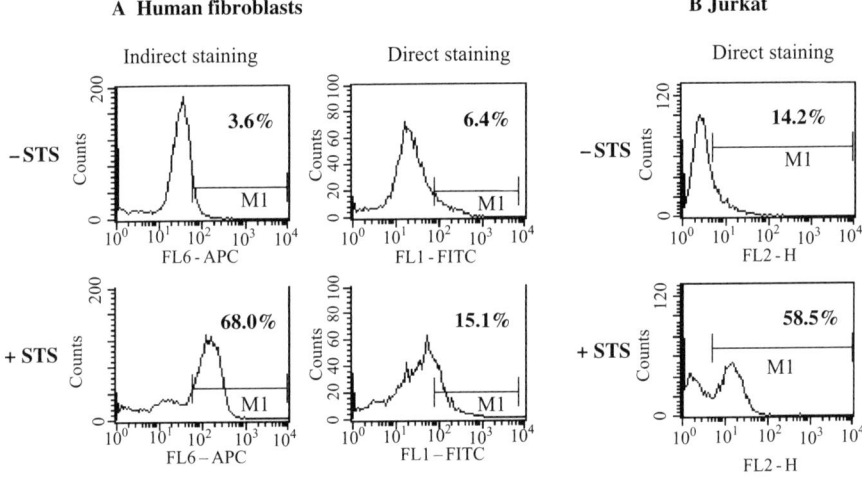

Fig. 2. Flow cytometric detection of activated caspase-3 after indirect and direct staining. (**A**) Human fibroblasts either treated with 1 μM staurosporine (+STS) for 10 h or left untreated (–STS) were stained for active caspase-3 using the indirect method with the rabbit anti-active caspase-3 antibody and allophycocyanin (APC)-conjugated anti-rabbit antibody (left panels) or the direct staining method with the FITC-conjugated anti-active caspase-3 antibody (right panels). (**B**) Jurkat cells either treated with 1 μM staurosporine (+STS) for 3 h or left untreated (–STS) were stained for active caspase-3 using the direct method with the PE-conjugated anti-caspase-3 antibody. The percentage of apoptotic cells is indicated in the upper right corner.

3. Incubate for 30–60 min at room temperature in the dark.
4. Add 1 ml 0.1% Tween solution.
5. Centrifuge 300 g for 5 min at room temperature; discard the solution.
6. Add 200–400 µl staining solution and transfer to a 5-ml round-bottom tube.
7. Analyze the cells by flow cytometry (*see* **Note 8**). The analysis will be done in FL1 channel for FITC or FL2 channel for PE (*see* **Fig. 2** for an example of expected results).

4. Notes

1. In this report, we are describing two alternate protocols for the detection of activated caspase-3 by flow cytometry. The first procedure described in **Subheading 3.2** gives the advantage of being able to choose a fluorophore that is not FITC or PE. In this report, we chose APC as an alternate fluorophore. This is particularly useful if we are studying other cell marker(s) in addition to activated caspase-3 and are required to use multiple fluorophores. The second method in **Subheading 3.3** is a one-step labeling protocol as FITC- or PE-conjugated antibody is used. Both methods are suitable for the simultaneous study of other cellular markers either surface or internal molecules. In the case of the study of a surface maker in addition to active caspase-3, the cell will have to first be stained for the cell surface molecule prior to proceeding with one of the two protocols. Additionally, in some cases, one protocol gives a better staining results than the other one, as demonstrated in **Fig. 2** for human fibroblasts. These protocols have been successfully adapted and applied to the measure of T-cell-mediated cytotoxicity *(17–19)*.
2. Pure preparations of paraformaldehyde are recommended for best results. Formic acid may be generated from formaldehyde by exposure to excess heat and light and can be present in some commercially available solutions that also can contain methanol as a stabilizing agent if not buffered. Both formic acid and methanol can have a negative effect on the structures of intracellular antigens. Paraformaldehyde solution stored at 4°C in the dark is stable for about 1 month as long as a pH of approximately 7.2 is maintained.
3. For cells in suspension, collect the cells by centrifugation (300 g for 5 min) and wash the cell pellet with PBS. For adherent cells, either trypsinize or scrape the cells into the medium using a rubber policeman, collect the cells by centrifugation (300 g for 5 min), and wash the cell pellet with PBS.
4. The entire procedure can be performed in 5-ml polypropylene round-bottom tube.
5. The cells can be stored overnight at 4°C at this step after centrifugation (300 g 5 min) and resuspension of the cells in PBS. No negative effect or change in the labeling was observed when doing so.
6. We have determined for our cell line (primary human fibroblasts) that the minimum amount of anti-active caspase-3 antibody we could use and still retain a good staining is 0.5 µl (0.25 µg) with a 30-min incubation. The amount of anti-active

caspase-3 antibody was decreased to 0.25 µl (0.125 µg) with the incubation time increased to 45 min.

7. This step was recommended by the manufacturer of the secondary antibody (Jackson ImmunoResearch Laboratories) to decrease potential background. It was omitted for several experiments without any obvious increase of background.

8. If needed, the cells could be stored overnight (up to 16 h) at 4°C in the dark in the staining solution before to be analyzed by the flow cytometry.

9. We have used 10 µl those antibodies with 5×10^5 to 1×10^6 cells with good success. However, the manufacturer recommends using 20 µl/1×10^6 cells.

References

1. Hengartner, M. O., and Horvitz, H. R. (1994) Programmed cell death in Caenorhabditis elegans. *Curr Opin Genet Dev* **4**, 581–6.

2. Nagata, S. (1997) Apoptosis by death factor. *Cell* **88**, 355–65.

3. Ashkenazi, A., and Dixit, V. M. (1999) Apoptosis control by death and decoy receptors. *Curr Opin Cell Biol* **11**, 255–60.

4. Beauvais, F., Michel, L., and Dubertret, L. (1995) Human eosinophils in culture undergo a striking and rapid shrinkage during apoptosis. Role of K+ channels. *J Leukoc Biol* **57**, 851–5.

5. Maeno, E., Ishizaki, Y., Kanaseki, T., Hazama, A., and Okada, Y. (2000) Normotonic cell shrinkage because of disordered volume regulation is an early prerequisite to apoptosis. *Proc Natl Acad Sci USA* **97**, 9487–92.

6. Liu, X., Li, P., Widlak, P., Zou, H., Luo, X., Garrard, W. T., and Wang, X. (1998) The 40-kDa subunit of DNA fragmentation factor induces DNA fragmentation and chromatin condensation during apoptosis. *Proc Natl Acad Sci USA* **95**, 8461–6.

7. Takahashi, A., Alnemri, E. S., Lazebnik, Y. A., Fernandes-Alnemri, T., Litwack, G., Moir, R. D., Goldman, R. D., Poirier, G. G., Kaufmann, S. H., and Earnshaw, W. C. (1996) Cleavage of lamin A by Mch2 alpha but not CPP32: multiple interleukin 1 beta-converting enzyme-related proteases with distinct substrate recognition properties are active in apoptosis. *Proc Natl Acad Sci USA* **93**, 8395–400.

8. Brown, D. G., Sun, X. M., and Cohen, G. M. (1993) Dexamethasone-induced apoptosis involves cleavage of DNA to large fragments prior to internucleosomal fragmentation. *J Biol Chem* **268**, 3037–9.

9. Sun, X. M., and Cohen, G. M. (1994) Mg(2+)-dependent cleavage of DNA into kilobase pair fragments is responsible for the initial degradation of DNA in apoptosis. *J Biol Chem* **269**, 14857–60.

10. van den Eijnde, S. M., Boshart, L., Baehrecke, E. H., De Zeeuw, C. I., Reutelingsperger, C. P., and Vermeij-Keers, C. (1998) Cell surface exposure of phosphatidylserine during apoptosis is phylogenetically conserved. *Apoptosis* **3**, 9–16.

11. Green, D. R., and Reed, J. C. (1998) Mitochondria and apoptosis. *Science* **281**, 1309–12.

12. Medema, J. P., Scaffidi, C., Kischkel, F. C., Shevchenko, A., Mann, M., Krammer, P. H., and Peter, M. E. (1997) FLICE is activated by association with the CD95 death-inducing signaling complex (DISC). *EMBO J* **16**, 2794–804.

13. Chinnaiyan, A. M. (1999) The apoptosome: heart and soul of the cell death machine. *Neoplasia* **1**, 5–15.

14. Sebbagh, M., Renvoize, C., Hamelin, J., Riche, N., Bertoglio, J., and Breard, J. (2001) Caspase-3-mediated cleavage of ROCK I induces MLC phosphorylation and apoptotic membrane blebbing. *Nat Cell Biol* **3**, 346–52.

15. Coleman, M. L., Sahai, E. A., Yeo, M., Bosch, M., Dewar, A., and Olson, M. F. (2001) Membrane blebbing during apoptosis results from caspase-mediated activation of ROCK I. *Nat Cell Biol* **3**, 339–45.

16. Leverrier, Y., and Ridley, A. J. (2001) Apoptosis: caspases orchestrate the ROCK 'n' bleb. *Nat Cell Biol* **3**, E91–3.

17. Jerome, K. R., Sloan, D. D., and Aubert, M. (2003) Measurement of CTL-induced cytotoxicity: the caspase 3 assay. *Apoptosis* **8**, 563–71.

18. Jerome, K. R., Sloan, D. D., and Aubert, M. (2003) Measuring T-cell-mediated cytotoxicity using antibody to activated caspase 3. *Nat Med* **9**, 4–5.

19. Aubert, M., and Jerome, K. R. (2003) Apoptosis prevention as a mechanism of immune evasion. *Int Rev Immunol* **22**, 361–71.

20. Yonehara, S., Ishii, A., and Yonehara, M. (1989) A cell-killing monoclonal antibody (anti-Fas) to a cell surface antigen co-downregulated with the receptor of tumor necrosis factor. *J Exp Med* **169**, 1747–56.

21. Tartaglia, L. A., Ayres, T. M., Wong, G. H., and Goeddel, D. V. (1993) A novel domain within the 55 kd TNF receptor signals cell death. *Cell* **74**, 845–53.

22. Wiley, S. R., Schooley, K., Smolak, P. J., Din, W. S., Huang, C. P., Nicholl, J. K., Sutherland, G. R., Smith, T. D., Rauch, C., Smith, C. A., and et al. (1995) Identification and characterization of a new member of the TNF family that induces apoptosis. *Immunity* **3**, 673–82.

23. Boesen-de Cock, J. G., Tepper, A. D., de Vries, E., van Blitterswijk, W. J., and Borst, J. (1999) Common regulation of apoptosis signaling induced by CD95 and the DNA-damaging stimuli etoposide and gamma-radiation downstream from caspase-8 activation. *J Biol Chem* **274**, 14255–61.

24. Furlong, I. J., Lopez Mediavilla, C., Ascaso, R., Lopez Rivas, A., and Collins, M. K. (1998) Induction of apoptosis by valinomycin: mitochondrial permeability transition causes intracellular acidification. *Cell Death Differ* **5**, 214–21.

25. Moreno, M. B., Memon, S. A., and Zacharchuk, C. M. (1996) Apoptosis signaling pathways in normal T cells: differential activity of Bcl-2 and IL-1beta-converting enzyme family protease inhibitors on glucocorticoid- and Fas-mediated cytotoxicity. *J Immunol* **157**, 3845–9.

26. Kruman, I., Guo, Q., and Mattson, M. P. (1998) Calcium and reactive oxygen species mediate staurosporine-induced mitochondrial dysfunction and apoptosis in PC12 cells. *J Neurosci Res* **51,** 293–308.
27. Gamen, S., Anel, A., Perez-Galan, P., Lasierra, P., Johnson, D., Pineiro, A., and Naval, J. (2000) Doxorubicin treatment activates a Z-VAD-sensitive caspase, which causes deltapsim loss, caspase-9 activity, and apoptosis in Jurkat cells. *Exp Cell Res* **258,** 223–35.

6

The ChemoFx® Assay: An Ex Vivo Chemosensitivity and Resistance Assay for Predicting Patient Response to Cancer Chemotherapy

Stacey L. Brower, Jeffrey E. Fensterer, and Jason E. Bush

Summary

The ChemoFx® Assay is an ex vivo assay designed to predict the sensitivity and resistance of a given patient's solid tumor to a variety of chemotherapy agents. A portion of a patient's solid tumor, as small as a core biopsy, is mechanically disaggregated and established in primary culture where malignant epithelial cells migrate out of tumor explants to form a monolayer. Cultures are verified as epithelial and exposed to increasing doses of selected chemotherapeutic agents. The number of live cells remaining post-treatment is enumerated microscopically using automated cell-counting software. The resultant cell counts in treated wells are compared with those in untreated control wells to generate a dose-response curve for each chemotherapeutic agent tested on a given patient specimen. Features of each dose-response curve are used to score a tumor's response to each ex vivo treatment as "responsive," "intermediate response," or "non-responsive." Collectively, these scores are used to assist an oncologist in making treatment decisions.

Key Words: Chemosensitivity and resistance assay (CSRA); primary tissue culture; personalized medicine; cancer; ex vivo chemoresponse.

1. Introduction

Unfortunately, one in three Americans will develop cancer during their lifetime. Although many advances have been made in detecting, diagnosing, and treating cancer, chemotherapy (the primary means of treatment) is only effective in approximately one-fourth of those treated. The use of ineffective

From: *Methods in Molecular Biology, vol. 414: Apoptosis and Cancer*
Edited by: G. Mor and A. B. Alvero © Humana Press Inc., Totowa, NJ

chemotherapy can lead to unnecessary toxicity and costs, delay of more effective treatment, and the potential for the development of cross-resistance to additional therapies. The ability to individualize therapy by providing the treating oncologist with ex vivo response information on a panel of chemotherapeutics agents aids in selecting between therapeutically equivalent treatment options and provides hope for more effective therapy and improved outcomes *(1)*.

The ChemoFx® Assay is an ex vivo assay in which the tumor cells, excised from individual cancer patients and grown in primary culture as a monolayer, are challenged with a variety of chemotherapeutic agents over a range of concentrations *(2,3,4,5,6)*. The ChemoFx® Assay is predicated on the biological phenomenon that, for cells that grow adherent in culture as a monolayer, cells lose their adherent qualities and lift off from the culture surface when they die (*see* **Note 1**). Resistance to chemotherapy cannot be predicted by either clinical or histological examination. Historically, the ex vivo sensitivity and resistance of tumor cells has been evaluated as a tool for predicting the clinical response of the patient to therapy *(7)*. Chemoresponse assays that have been developed to determine sensitivity and/or resistance of tumor cells can be grouped into three broad categories: (i) indirect assays that measure cellular metabolic activity (e.g., through ATP utilization) *(8)*, (ii) indirect assays that measure incorporation of a radioactive precursor (such as ^{3}H-thymidine) *(9)*, and (iii) assays that quantify cell effects by direct visualization of cells following exposure to the anti-cancer agents. The ChemoFx® Assay is an example of the third category of assays.

Although there are other chemosensitivity and resistance assays (CSRAs) that have the same underlying premise of predicting patient response to chemotherapy *(10,11)*, the ChemoFx® Assay offers many advantages over these commercial and research-based assays (*see* **Table 1**). Notably the ChemoFx® Assay requires a minimum of 35 mg of tissue for testing; this is approximately a 30-fold reduction over the minimum requirement of other CSRAs and allows for testing of tissue obtained from less-invasive procedures such as vacuum-assisted and core needle biopsies. The ChemoFx® Assay utilizes a proprietary cell-culture process to ensure the isolation and growth of malignant, epithelial cells, thereby ensuring that the assay results pertain only to the cells of interest (cancer cells) and not to normal, non-malignant cell types. Finally, the ChemoFx® Assay is able to provide the prescribing oncologist with information regarding the tumor's sensitivity *and* resistance to the agents that have been tested. This is accomplished through the use of a wide range of drug concentrations which allows the generation of a full dose-response curve. Other

Table 1
Comparison of Other Chemosensitivity and Resistance Assays (CSRAs)
(Past and Present) with The ChemoFx® Assay

Other CSRAs	The ChemoFx® Assay
1–2 g required	35 mg required
Test all cells received	Test epithelial cells
Labor-intensive, highly variable	Automated, reproducible
Resistance only	Sensitivity and resistance
Single-point threshold	Dose-response curve
Low yield	High accessibility (>90%)
Long turn around time	2–3 week turn around time
Single-drug therapies	Combination therapies
Non-proprietary	Patent protected
Low amounts of clinical data supporting claims	Retrospective and on-going prospective trials to further demonstrate claim

CSRAs use only one or a few drug concentrations, and they can only comment on drug resistance, and not sensitivity.

2. Materials

2.1. Tissue Culture

1. Tissue culture media:

   a. McCoy's 5A (Mediatech, Herndon, VA).
   b. RPMI 1640 (Mediatech).
   c. Mammary epithelial growth media (MEGM; Cambrex, Walkersville, MD).
   d. Bronchial epithelial growth media (BEGM; Cambrex).

2. Fetal bovine serum (FBS; Hyclone, Logan, UT).
3. Hanks' balanced salt solution (HBSS; Mediatech).
4. Trypsin/EDTA (Mediatech).
5. Antibiotics/antimycotics:
   a. Penicillin–streptomycin solution (Mediatech).
   b. Gentamicin reagent solution (Invitrogen Corporation, Grand Island, NY).
   c. Fungisone (Invitrogen Corporation).
   d. Cipro® I.V. (ciprofloxacin) (Oncology Therapeutics Network, South San Francisco, CA).
   e. Nystatin (Sigma-Aldrich, St. Louis, MO).

6. PureCol™ (Inamed Biomaterials, Fremont, CA).
7. Cell-culture flasks (Greiner, Falcon, and PGC Scientifics, Frederick, MD).
8. Sterile scalpels (#10 and #11 blades).

2.2. Immunocytochemistry

1. Mouse monoclonal antibodies:

   a. Negative control mouse IgG_1 and IgG_{2a} (DakoCytomation, Carpinteria, CA).
   b. Anti-cytokeratin 20 (DakoCytomation).
   c. Anti-cytokeratin (CAM 5.2) (BectonDickinson, San Jose, CA).
   d. Anti-cytokeratin 7 (BioGenex, San Ramon, CA).
   e. Anti-cytokeratin AE1/AE3 (Chemicon, Temecula, CA).
   f. Anti-fibroblast antigen (Calbiochem, San Diego, CA).

2. Goat anti-mouse antibody conjugated to Alexa Fluor® 488 (Molecular Probes, Carlsbad, CA).
3. Tris-buffered saline (TBS; Sigma).
4. Methanol (VWR International, West Chester, PA).
5. Terasaki 60-well microtiter plates (PGC Scientifics).

2.3. Assay

1. 384-well microtiter plates (Costar and VWR International).
2. Deep-well basins (DOT Scientific, Burton, MI).
3. 2.2 ml 96-deep-well plates (VWR International).
4. Anti-cancer agents (Oncology Therapeutics Network).
5. Ethanol, anhydrous, 95% fixing grade (VWR International).
6. 4′,6-diamidino-2-phenylindole (DAPI; Molecular Probes).

2.4. Equipment

1. Class IIB laminar flow hoods.
2. CO_2 incubators.
3. Fluorescence microscope (Carl Zeiss MicroImaging, Thornwood, NY, USA).
4. Image analysis microscope and software (Olympus, American, Center Valley, PA, USA).
5. Oasis LM1200 liquid handler (Dynamic Devices, Newark, DE).

3. Methods

The methods described herein detail the process of the ChemoFx® Assay from the time of receipt of the specimen through the interpretation of assay results. The ChemoFx® Assay is applicable to solid tumor specimens; fluid specimens (ascites and pleural fluid) are also acceptable (*see* **Note 2**).

Specimens submitted for testing should be representative of the tissue submitted for pathologic examination, placed immediately into media, and shipped overnight (*see* **Note 3**).

3.1. Accessioning, Explanting, and Initiating Culture

Upon receipt, the specimen is initially examined for size/volume, potential contamination, and other indications of proper shipping conditions such as temperature and number of days in transit. In addition, a pathology report from the requesting institution must be sent as soon after receipt of the specimen as possible. The pathology report is necessary to confirm a malignant diagnosis and the tissue of origin of the tumor.

3.1.1. Processing Solid Tumor Specimens

1. Pour the shipping media from the specimen bottle and evenly distribute into two 50-ml conical tubes, leaving the solid tumor in the bottle.
2. Centrifuge the shipping media at $800 \times g$ for 3 min.
3. Carefully decant and discard the supernatant without disturbing the cell pellet. This may be done through pouring or using a pipet.
4. Resuspend the cell pellets in an appropriate volume of cell culture media and transfer to the appropriate-sized culture flask (*see* **Notes 4–6**).
5. Transfer the pellet flask to the incubator (37°C, 5% CO2) for maintenance.
6. Under a laminar flow hood, remove the solid tumor from the specimen bottle and transfer to a sterile culture dish. If needed, sterile forceps may be used to assist (*see* **Note 7**).
7. Using disposable, sterile scalpels, mince the tumor into the smallest explants possible (ideally 1 mm^3 pieces).
8. Based on the size of the tumor, determine how many culture flasks will be needed (*see* **Notes 8**).
9. Using a 25 ml serological pipet, transfer the appropriate volume of the appropriate type of media to the culture dish containing tumor explants (*see* **Notes 9** and **10**).
10. Using a swirling motion, create a slurry of media and tumor explants, then evenly transfer to the desired culture flasks.
11. Swirl to evenly distribute the explants on the bottom of each flask.
12. Tilt each flask at a slight angle for 10–20 min, allowing the media to pool at the bottom edge while leaving the tissue explants to adhere to the growth surface of the flask. Do not allow explants to dry out.
13. Carefully return each flask to a horizontal position so that the adhered explants are not dislodged.
14. Transfer all explant flasks to the incubator (37°C, 5% CO_2) for maintenance.

3.1.2. Processing Fluid Specimens

1. Swirl and/or invert the fluid specimen container to resuspend the cells into solution.
2. Evenly distribute the entire contents of the container into two conical tubes (either 50 or 200 ml, depending on volume of specimen).
3. Centrifuge the fluid specimen at $400 \times g$ for 10 min.
4. Carefully decant and discard the supernatant without disturbing the cell pellet. This may be done through pouring or using a pipet.
5. Resuspend the cell pellets in an appropriate volume of cell culture media and transfer to the appropriate size culture flasks (*see* **Notes 6** and **11**).

3.2. Monitoring and Maintaining Cultures

3.2.1. Examination of Cultures

1. Inspect media in culture flasks for color. All media used in the assay contain phenol red as an indicator of pH. Neutral pH results in a red color. If media have a yellow (acidic pH) or pink (alkaline pH) color, the media should be changed (*see* **Note 12**).
2. Inspect media in culture flasks for turbidity or cloudiness. Turbid culture medium is an indication of microbial contamination.
3. Under at least $\times 20$ magnification, inspect more closely for contaminants such as bacteria, yeast, or fungus (*see* **Note 13**). If contamination is suspected, the culture may be placed in antibiotic cell culture media (*see* **Note 7**).
4. Inspect the growth surface of each flask for cell growth. Gently slide flask back and forth to determine whether cells have adhered to the bottom of the flask.
5. Determine and record the percentage of the total culture surface area (confluency) that is covered with adherent cells.
6. Observe the morphology of cells adherent to the flask. Note whether the cells appear to be fibroblastic (elongated, spindle-shaped, densely-packed "swirls") or epithelial in nature.

3.2.2. Flask Actions

Once all flasks for each culture have been read, there are a variety of actions that may result.

1. Regardless of the color and quality of the cell culture media, it should be changed (decanted and replaced with an equal volume of fresh media) at least once a week.
2. If little or no growth is seen in a flask, the percentage of FBS in the media may be increased (e.g., from 2 to 5% FBS).
3. If there is a mixed population of fibroblast and epithelial cells, the flask should undergo differential trypsinization (*see* **Note 14**).
4. If enough epithelial cells are present (*see* **Note 15**), the culture can be prepared for plating for the assay.

3.3. Plating Cultures

3.3.1. Trypsinization and Counting

1. Examine cultures for confluency and the presence of debris and/or contamination and general health of the cells.
2. Pour off the growth medium and any remaining explants and rinse each flask at least once with HBSS (without calcium and magnesium) to rinse the cell monolayer and remove any debris.
3. Pipet an appropriate amount of trypsin (0.25% with EDTA) into each flask (*see* **Note 16**) and swirl to ensure complete coverage.
4. Monitor for cell detachment under the microscope. If cells are not lifting off, gently strike the side of the flask. If still no detachment, place the flask in the incubator (37°C, 5% CO_2) for 30 s to 1 min. Continue cycles of striking the flask and placing it in the incubator until the cells detach. In between each step, monitor the flask under the microscope.
5. Once all cells have detached, add trypsin neutralizer to the flask (approximately twice the volume of trypsin that had been added to that flask). Using a pipet, rinse the flask several times.
6. Transfer the cell suspension to a 15 ml conical tube and centrifuge at 800 × *g* for 3 min.
7. Discard the supernatant carefully so as not to disturb the cell pellet.
8. Resuspend the cell pellet in 1.0 ml appropriate growth medium.
9. Remove 30 μl well-mixed cell suspension and transfer to 1 well of a 96-well plate. Add 30 μl trypan blue and mix thoroughly.
10. Fill both sides of hemacytometer with 10 μl cell suspension/trypan blue mixture and transfer to the microscope stage.
11. Count the number of viable (clear) cells in each of five squares on each side of the hemacytometer (10 squares total).
12. The cell concentration is determined using the following equation:

$$\frac{\text{Total number of viable cells in 10 squares}}{10} \times 2 \times 10,000 = \text{Number of cells/ml}$$

3.3.2. Plating Cells for the Assay

1. Dilute the cell suspension to obtain a concentration of 40,000 cells/ml.
2. Dispense 20 μl of 40,000 cells/ml suspension into each of at least four wells of a Terasaki plate. Place plate in incubator for 24 h. This plate will be used for immunocytochemistry (*see* **Subheading 3.4.**).
3. From the 40,000 cells/ml suspension, further dilute to create an 8000 cells/ml cell suspension.
4. After thorough mixing, transfer the cell suspension to a sterile eight-row basin and place it in the appropriate position on the liquid handler deck. Also place an empty 384-well plate(s) on the deck.

5. After selecting the appropriate plating program, dispense 40 µl cell suspension into each well of a 384-well plate using the liquid handler. For each drug treatment, a total of 33 wells will be used (10 drug concentrations and 1 untreated control, all done in triplicate).
6. Transfer the plates containing cells to the incubator (37°C, 5% CO_2) for 24 h to allow the cells to attach to the bottom of each well before treatment.

3.4. Immunocytochemistry

The malignant cell type expressed in solid tumor adenocarcinomas is the epithelial cell. The other major cell type present in solid tumors is the non-malignant stromal fibroblast. Morphologically, epithelial cells can be of almost any size and shape, whereas fibroblasts are generally elongate, spindle-shaped, and often grow in "swirls." As cell morphology alone cannot definitively distinguish cell type, fluorescence-based staining for cytokeratins has been used as a marker to identify epithelial cells (fibroblasts do not express cytokeratins) (*see* **Fig. 1**).

3.4.1. Fluorescence Staining of Cultures

At the time of plating for the assay, trypsinized cultures are also plated for immunocytochemistry analysis (*see* **Subheading 3.3.2.**).

1. After a 24 h incubation period to allow for complete cell attachment, retrieve immunocytochemistry (Terasaki) plates from the incubator and rinse by flooding the plates with HBSS. This is done carefully so as not to disrupt the cells from the plate surface.

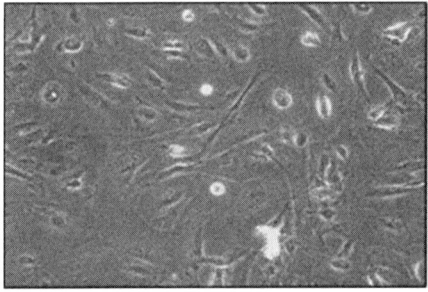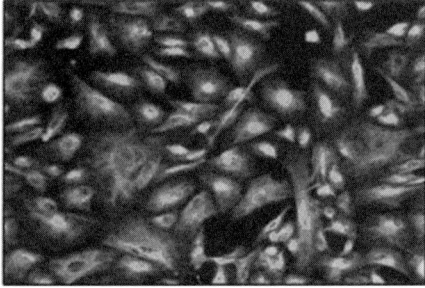

Fig. 1. Representative epithelial cell staining as determined by immunocytochemistry. Phase-contrast microscopy (left panel) shows the morphology of the cells being grown. Fluorescence microscopy (right panel) shows epithelial cell detection after immunofluorescence staining for detection of the cytokeratins.

2. Gently pour off HBSS and "fix" the cells by flooding each plate with 100% methanol (which has been stored at –20°C) for 10 min.

3. Gently pour off methanol and flood each plate with TBS to rinse off excess methanol. Then pour off TBS, invert plate on absorbent pad, and strike plate gently twice against the pad to remove excess TBS from the wells.

4. To the appropriate wells, add 10 µl each diluted primary antibody (*see* **Note 17**). Incubate at room temperature for 60 min.

5. After primary antibody incubation, rinse each plate thrice by flooding with TBS, agitating briefly, and pouring TBS off. After all of the rinses have been completed, pour off the TBS and then invert the plates and strike gently twice on an absorbent pad to remove excess TBS from the wells.

6. Add 10 µl diluted secondary antibody (goat anti-mouse IgG conjugated to Alexa Fluor® 488) to each well that had primary antibody. Incubate for 30–60 min at room temperature in the dark. The fluorescent tag on the secondary antibody is light sensitive.

7. After secondary antibody incubation, rinse each plate thrice by flooding with TBS, agitating briefly, and pouring TBS off. After all the rinses have been completed, pour off the TBS and then invert the plates and strike gently twice on an absorbent paper towel to remove excess TBS from the wells.

8. Flood each plate with DAPI (diluted 1:1000 in TBS) for 10 min at room temperature in the dark. DAPI is light sensitive.

9. Pour off DAPI solution and quickly rinse twice with TBS or tap water.

10. Flood each plate with TBS and store at 4°C until viewing under microscope.

3.4.2. Reading and Interpreting Results

1. Under ×10 magnification, view the stained wells for each specimen under phase and fluorescence to approximate the percentage of cells that stain positively for each primary antibody or cocktail of antibodies (*see* **Note 18**).

2. If at least 65% of the cells in a given culture stain positively for cytokeratins, then that specimen is deemed suitable to testing in the assay.

3. If less than 65% of the cells stain positively for cytokeratins, then that specimen is not appropriate for use in the assay, and it is terminated (*see* **Note 19**).

3.5. Preparing, Applying, and Removing Anti-Cancer Agents

3.5.1. Preparation of Anti-Cancer Agents

Anti-cancer agents are supplied in a variety of forms. For this application, lyophilized powder and liquid drug are most commonly used. Lyophilized powder agents are reconstituted to a concentrated form using sterile technique with the diluent recommended by the manufacturer; liquid drugs are already in their concentrate form. Never using more than a 1:50 dilution, drug concentrates go through a series of dilutions to bring them to the highest drug dose (i.e., dose 10) used in the assay. For higher throughput applications, a large batch (50 ml) of dose 10 may be prepared and stored at –80°C in aliquots of 1.5 ml each. The length of time of storage at –80°C without compromised drug activity must be individually determined for each anti-cancer agent.

3.5.2. Dilution and Application of Anti-Cancer Agents

1. Obtain a dose 10 frozen aliquot of each agent needed from the –80°C freezer and thaw at room temperature.
2. Under a laminar flow hood, aliquot 1.5 ml of each dose 10 into a separate well in column 11 of a 96-deep-well plate.
3. Place a sterile cover over the deep-well plate and transfer to the liquid handler, placing the plate in its appropriate location.
4. After selecting the appropriate dilution program, dispense the appropriate volume of media into wells 1–10 in each row of the deep well plate that contains a dose 10 aliquot using the liquid handler.
5. Then perform serial dilutions of each drug using the appropriate dilution factor and adequately mixing each dose throughout the process by the liquid handler. In total, 10 drug doses are created. The 11th well contains only media for control well treatment.
6. After all drugs have been diluted (*see* **Note 20**), using a multichannel pipet tip head, aspirate an appropriate volume of each dose of each drug and then dispense 40 μl each dose into each replicate of the patient cells that have been plated using the liquid handler. This yields a total volume of 80 μl in each well of the 384-well plate (*see* **Fig. 2**).
7. Once all drugs have been applied, cover the 384-well plate with its sterile lid, remove from the liquid handler, and place in an incubator (37°C, 5% CO_2) for 72 h.

3.5.3. Removal of Anti-Cancer Agents

1. Retrieve the appropriate plate from the incubator and place in the appropriate position on the deck of the liquid handler. Remove the lid of the plate.
2. Fill one deep-well basin approximately half full with HBSS, and fill a second one approximately half full with 100% ethanol.
3. Place basins in the appropriate positions on the deck of the liquid handler.
4. After selecting the appropriate fixing program, remove the anti-cancer agents using the liquid handler and then add 70 μl HBSS to each well.
5. Then remove the HBSS and add 70 μl ethanol to each well.

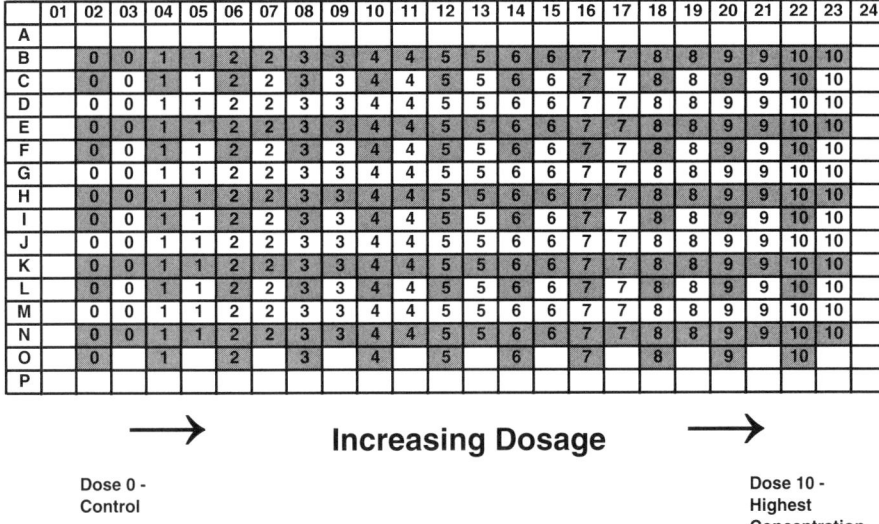

Fig. 2. Layout of a typical 384-well plate used in the ChemoFx® Assay. Each color represents a different agent or combination of agents. The dose 10 concentration range for each treatment has three replicates. In total, up to nine treatments can be included on one plate. Any unused wells are filled with Hanks' balanced salt solution (HBSS).

6. Fix the cells in the ethanol for at least 10 min. Plates can be stored in ethanol at room temperature for an extended period of time; additional ethanol should be added to the plates to prevent them from drying out. Remove the ethanol immediately before DAPI staining.

3.6. Staining and Automated Cell Counting

1. Fill one deep well basin approximately half full with DAPI (diluted 1:1000 in TBS), and fill a second one approximately half full with sterile water.
2. Place basins in the appropriate deck positions on the deck of the liquid handler.
3. Retrieve fixed 384-well plates and place them in the appropriate deck positions on the deck of the liquid handler. Remove plate lids.
4. After selecting the appropriate staining program, remove the ethanol using the liquid handler and then add 70 µl DAPI solution to each well.
5. Incubate with DAPI for at least 10 min. Plates can be stored with DAPI at 4°C for several days.

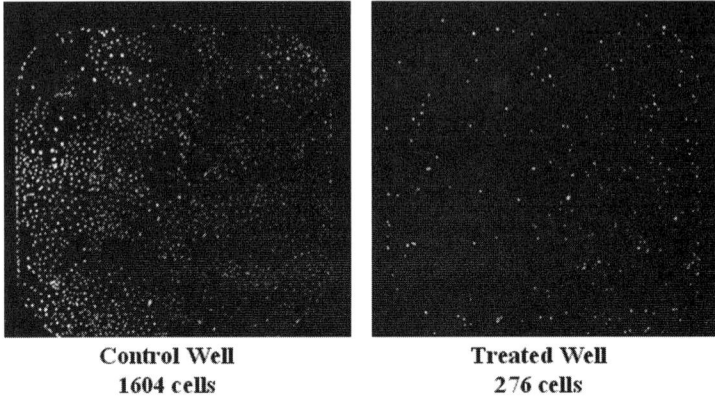

Control Well
1604 cells

Treated Well
276 cells

Fig. 3. Representative 4′,6-diamidino-2-phenylindole (DAPI)-stained control (left panel) and treated (right panel) wells following ex vivo chemotherapy treatment. The survival fraction (SF) is determined by direct cell counting:

$$SF = \frac{\text{Treated well cell counts}}{\text{Control well cell counts}} = \frac{276}{1604} = 0.17$$

6. Immediately before scanning, place plates on the liquid handler deck for automated removal of the DAPI solution and add 70 µl distilled water to each well.
7. Place each plate onto the deck of the scanning station. The scanning station consists of an inverted fluorescence microscope interfaced with a computer containing image analysis software. The scanning station also contains a camera equipped to capture and save acquired images.
8. Automatically scan each well of the plate. Parameters are optimized to detect, enumerate, and store the number of DAPI-stained nuclei in each well (*see* **Fig. 3**).

3.7. Interpreting Results

For each dose of a drug treatment and the corresponding untreated control wells for that treatment, the average cell counts from each drug dose is divided by the average cell counts from corresponding control wells. The resultant survival fraction (SF) represents the percentage of cells killed at each drug dose. The SF (*y*-axis) and drug dose (*x*-axis) are plotted against one another to form a dose-response curve that is unique to each specimen and drug treatment (*see* **Fig. 4**). The features of each dose-response curve are then used to score

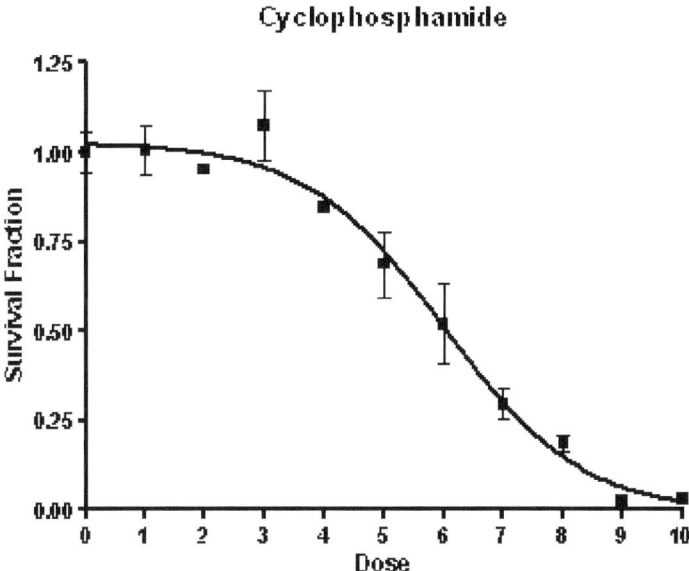

Fig. 4. Dose-response curve that represents the survival fraction (SF) at each drug dose compared with the corresponding untreated control wells.

a tumor's response to each ex vivo treatment as "responsive," "intermediate response," or "non-responsive." Collectively, these scores are used to assist an oncologist in making treatment decisions.

As with any biological model system, it is essential to understand and correct for (as much as possible) the inherent variability in and effect of perturbations on that system. Extensive studies have explored these features of the ChemoFx® Assay (*see* **Notes 21–23**).

4. Notes

1. Using a combination of CalceinAM (taken up by live cells) and ethidium homodimer-2 (taken up by dead cells) fluorescent dyes, it has been shown that less than 2% of the cells that have lifted off of the microtiter plate at the conclusion of the assay are alive.
2. The ChemoFx® Assay has not been optimized for use with leukemias and lymphomas.
3. Specimens must be shipped fresh and kept cool with 4°C cold packs. Frozen or fixed tissue cannot be used in the described assay.
4. Cell culture media is chosen according to the tumor type. For example, MEGM is used for breast specimens, and BEGM is used for lung specimens.

5. Typically, a T25 flask is adequate for the pellet. For this size flask, approximately 2.5 ml media should be used for resuspending the pellet. Alternatively, if the pellet is unusually small, resuspend in 1.0 ml media and establish in a T12.5 culture flask.

6. Culture flasks may be coated with PureCol™ to promote adherence to the culture surface.

7. Specimens that are potentially contaminated must be washed and initiated in culture using antibiotic/antimycotic media as described below:

    a. Cell pellets from shipping media are resuspended and incubated for 15 min (with rocking) with 5–20 ml antibiotic wash solution (150 U/ml penicillin–streptomycin, 40 μg/ml ciprofloxacin, 100 μg/ml gentamicin, 0.25 μg/ml amphotericin B, 100 U/ml nystatin in HBSS). Pellets are then centrifuged and established in culture using antibiotic cell culture medium (50 U/ml penicillin–streptomycin, 20 μg/ml ciprofloxacin, 80 μg/ml gentamicin, 100 U/ml nystatin in appropriate media) in place of regular growth medium for up to 3 days.

    b. After mincing solid specimens, resuspend explants in 5–20 ml antibiotic wash solution, transfer to a 50-ml conical tube, and incubate (with rocking) for 15 min. Centrifuge, discard supernatant, and resuspend in the appropriate volume antibiotic cell culture media. Follow same process for establishing explants in culture.

    c. Cultures should remain in antibiotic media for up to 3 days with no indication of contamination before transfer to regular media. The amount of time in antibiotic media should be limited as antibiotics can be toxic to the cultures.

8. Explants should cover 30–50% of the culture surface of each flask.

9. A 25 ml serological pipet works best because it has the widest mouth. In using smaller pipets, explants of a larger size may not be able to be transferred.

10. Suggested media volumes for explant attachment and tumor cell outgrowth are T12.5, 0.75–1.0 ml; T25, 1.5–2.0 ml; T75, 5.0–6.0 ml.

11. Suggested media volumes for flasking fluid specimens are T12.5, 2.0 ml; T25, 5.0 ml; T75, 12.0 ml. As the number of cells in the pellet of a fluid specimen is unknown, a combination of T25 and T75 flasks is advised.

12. Change in color of media may be indicative of the level of growth of the culture or the potential presence of contamination with microbial substances.

13. It is important to recognize the distinction between cellular debris and microbial contamination. Debris is dark and granular and moves through the media through Brownian movement. Bacterial contaminants are consistent in size and shape and are often motile.

14. Differential trypsinization is a technique that capitalizes on the different adherent properties of epithelial and fibroblast cells so that fibroblasts may be removed from the culture. Differential trypsinization is best performed immediately before plating for the assay.

a. Rinse the flask with media to remove any remaining explants (explants may be discarded or seeded into a new flask).

b. Add the appropriate volume of 0.25% trypsin/EDTA to the flask and observe under the microscope for detachment of fibroblast cells (fibroblasts are less adherent than epithelial cells and will detach more quickly).

c. When nearly all fibroblasts have lifted off (but before epithelial cells begin to), lightly strike the side of the flask to completely dislodge the fibroblasts.

d. Decant trypin/fibroblasts and proceed with trypsinization of the epithelial cells for hemacytometer counting (*see* **Subheading 3.3.1.**).

15. As a general rule, for T12.5 or T25 flasks, 10% confluency is needed for each anti-cancer agent to be tested. For T75, 5% confluency is needed per agent to be tested.

16. Suggested volumes for trypsin: T12.5, 1.0 ml; T25, 2.0 ml; T75, 4.0 ml.

17. A cocktail of negative control IgG antibodies (IgG_1 and IgG_{2a}) is used to account for any background staining. A cytokeratin cocktail containing AE1/AE3, CAM5.2, and CK7 is used to positively identify epithelial cells; CK20 is used to distinguish colon epithelial cells from all other organ types. The anti-fibroblast antibody is used to positively identify fibroblast (stromal) cells.

18. DAPI staining is used to assist in approximating the total number of cells in each well. This, in turn, assists in more accurately determining the percentage of cells (of the total that are attached in each well) that are staining positively for the antibody in question.

19. Using epithelial and fibroblast cell lines, it has been determined that cultures that contain more than 35% fibroblast cells have anti-cancer agent response profiles that differ from those of a culture that is purely epithelial cells. When cultures contain less than 35% fibroblast cells, chemoresponse is identical to that of pure epithelial cultures.

20. Anti-cancer agents differ in their stability profiles. Therefore, the least stable agents are diluted last and applied first, whereas the most stable agents are diluted first and applied last.

21. Process variability impacts the ability to observe biological differences. In developing an assay to assess the biological differences between tumor-derived cells, it is critical to ensure that the process variability is significantly less than the differences in biological signals. Several experiments were designed to determine the process variability. To capture the variability of the process and reduce biological variability as much as possible, HTB77 cell lines were used.

a. *Assessment of process variability*. Performing the assay under known conditions with several repetitions measures the repeatability of the process. In the experiment, three operators performed the assay as described in the text using carboplatin on HTB77 cell lines on six different days. **Figure 5** and **Table 2** exhibit the reproducibility of the biological assay. Of note is the difference

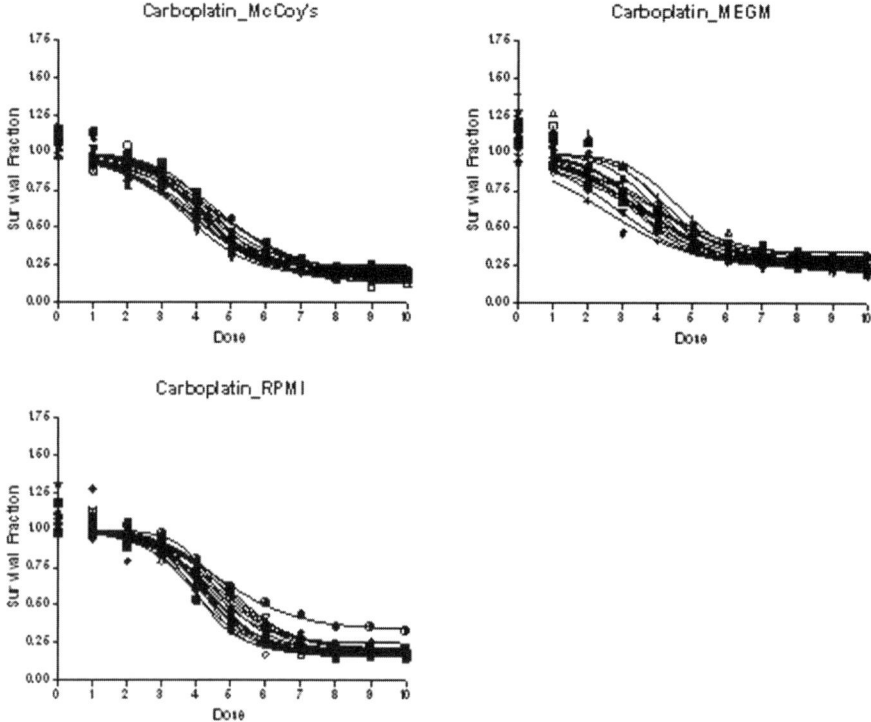

Fig. 5. Assessment of process variability. Data points are the mean of three replicates for each individual run of the process. Doses are ordinal with increasing number indicating higher concentration of the drug. Concentrations are identical between media types.

Table 2
Reproducibility of the ChemoFx® Assay Using HTB77 Cell Line

	Log EC50				
	Average	SD	Minimum	Maximum	CV (%)
Carboplatin–McCoy's	4.16	0.38	3.63	4.80	9.07
Carboplatin–MEGM	3.62	0.51	2.49	4.53	13.97
Carboplatin–RPMI	4.45	0.31	3.90	5.00	6.94

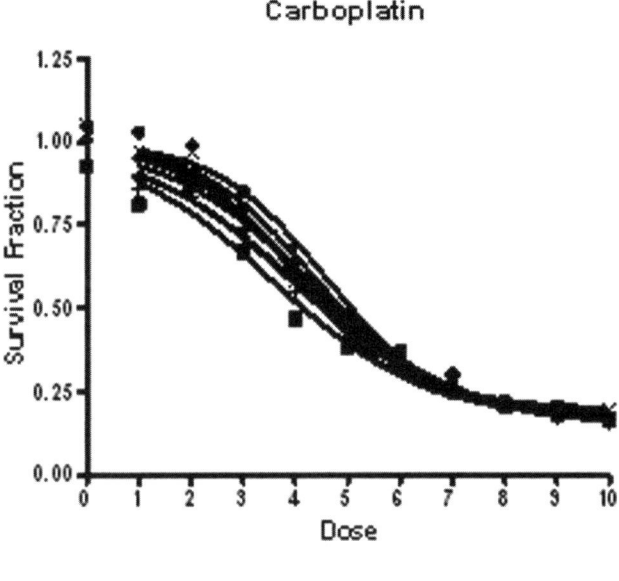

Fig. 6. Intra-operator process variability. Results were produced by a single operator on the same day using the same cell and drug suspensions.

in dose-response curves between media types. Once reproducibility is demonstrated, the results can serve as a quality control standard for the process. By comparing daily runs to expected outcomes, differences in drug preparation, deviations in cellular behavior, and errors in fixing and staining can be identified.

b. *Intra-operator process variability.* Similar to measuring the process variability across operators and days, the within-day process variability was measured for one operator. A single operator performed the assay eight times in a single day using the same HTB77 cells and the same preparation of carboplatin. Such reproducibility studies help to determine the extent of variability attributable to day-to-day and operator-to-operator factors (*see* **Fig. 6**).

c. *Inter-day drug stability.* Optimal use of a drug, logistical ease, and reduced variability can be achieved by preparing large quantities of the drug and freezing in smaller aliquots for use. The stability of the drug while frozen should be evaluated to ensure that changes in potency do not contribute to the variability of the process. Experiments were conducted measuring the potency of frozen drugs compared with freshly prepared drug over a course of several weeks. **Figure 7** depicts the response curves of carboplatin in RPMI-culture medium at the initial week, after 7 weeks frozen, and 14 weeks frozen. Intervening weeks were removed from the graph.

d. *Intra-day drug stability.* Process variability can be introduced by potency changes in the drug, and one significant source of degradation in potency is the instability of drugs in solution. As shown in **Fig. 8**, the stability of melphalan in RPMI medium degrades between 4 and 8 h. In the ChemoFx® Assay, the objective of measuring biological response differences between primary tumor tissues requires that the potency of the drug be equivalent throughout testing. Studies on drug stability in solution facilitate knowing the time limits associated with treating the drug after preparation.

e. *Sensitivity to timing of sequential events.* In a clinical laboratory process with defined tasks that must be accomplished on a daily basis, maintaining strict adherence to time windows can become a logistical problem. The assay described here is based on a 72 h exposure of the cells to chemotherapeutic agents after a 24 h incubation period where the cells adhere to the surface of the microtiter plate. It was found that cellular response profiles matched when the time between plating and treating ranged from 20 to 28 h (*see* **Fig. 9**). The findings support operational flexibility with respect to timing of key events.

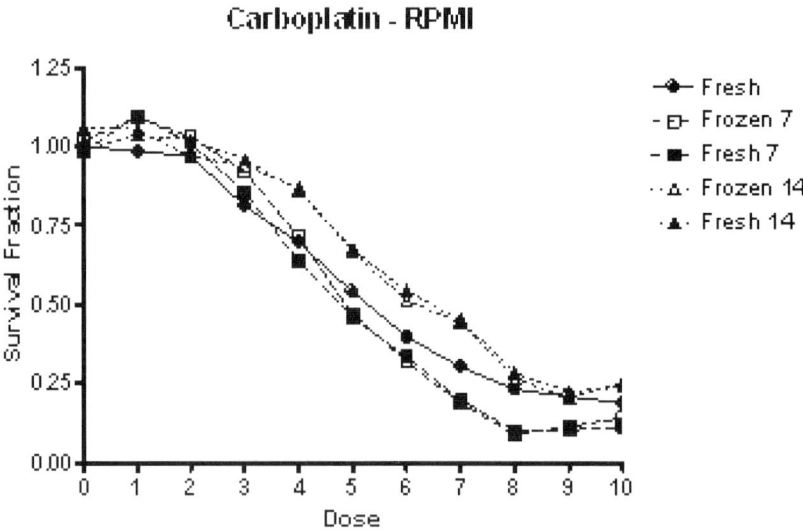

Fig. 7. Inter-day drug stability. Freezing does not affect drug potency in the assay.

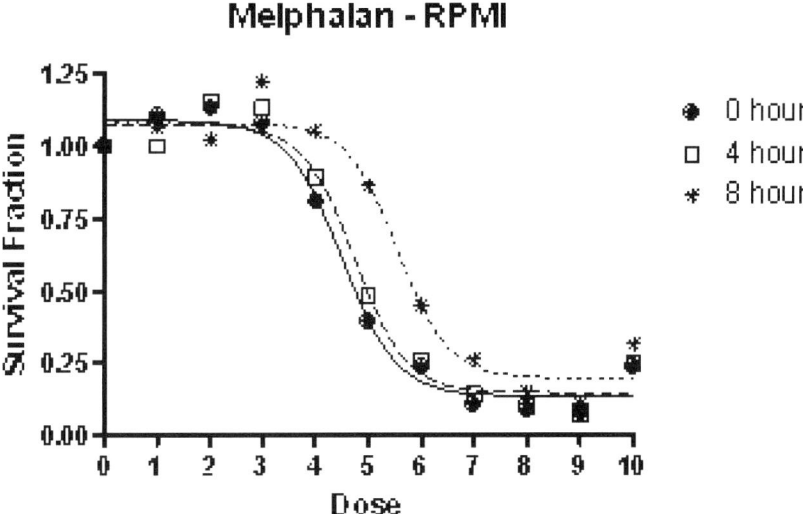

Fig. 8. Intra-day drug stability. The stability of drug in solution may decline over time.

22. Drug characteristics can impact the performance of a cell-based assay. As an example, cyclophosphamide degrades into two volatile components, a phosphoramide mustard responsible for the drug's anti-cancer properties and a vapor by-product, chloroethylaziridine (CEZ), which has been shown to contribute to cyclophosphamide's cytotoxic effect in vivo and in vitro *(12)*. In the presence

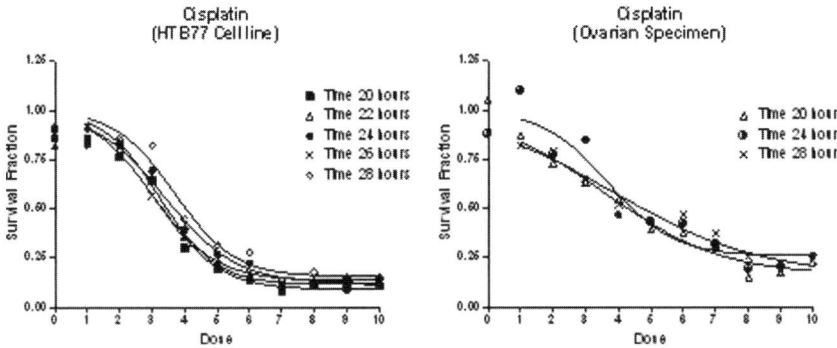

Fig. 9. Sensitivity of timing to sequential events. There is flexibility in the timing of operational events within the assay process.

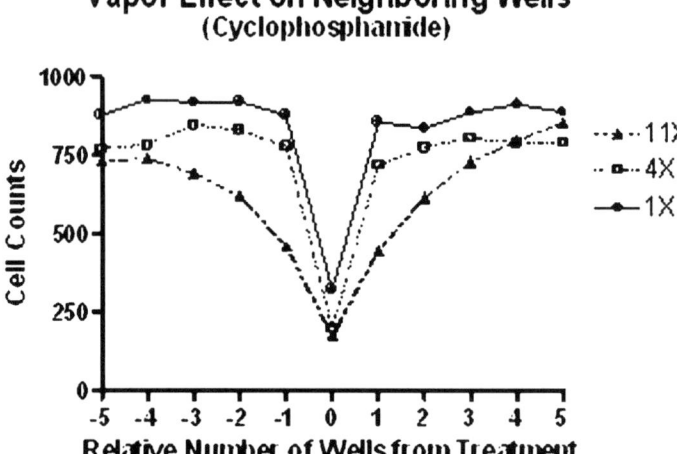

Fig. 10. Drug characteristics can impact the performance of a cell-based assay. A vapor [chloroethylaziridine (CEZ)] produced as a by-product of cyclophosphamide metabolism "spills over" to affect the viability of neighboring, untreated cells in a microtiter plate.

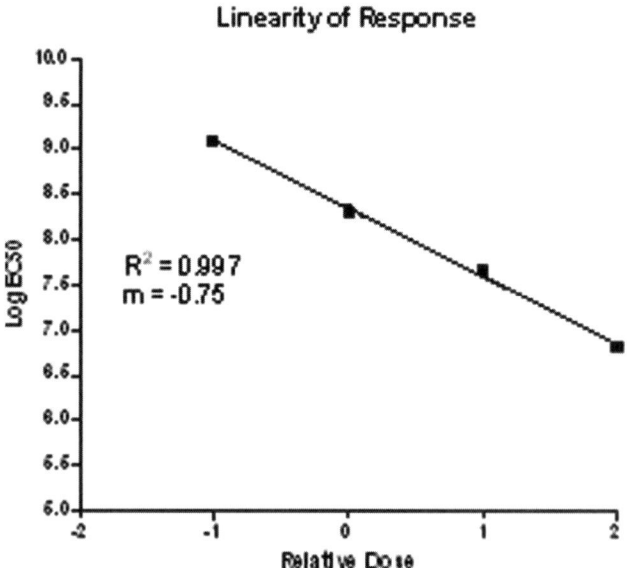

Fig. 11. Ability to discriminate perturbations of the assay. Alterations in drug concentration result in predictable shifts within the dose-response curve.

of the enzyme aldehyde dehydrogenase (ALDH), the vapor produced is minimal; however, some primary cells and immortalized cell lines have low levels of ALDH resulting in significant vapor, which can negatively impact neighboring wells' responses. **Figure 10** demonstrates the phenomenon of "spillover" of the vapor that occurs at higher concentrations relative to a $1\times$ reference.

23. Ability to discriminate perturbations of the assay. A good robustness test of a cell-based assay is to determine whether systemic changes can be caught and are predictable. In **Fig. 11**, the starting concentrations for the drug were increased or decreased by an entire dose. The linearity of the response demonstrates the ability of the assay to identify changes in potency of drug preparation.

Acknowledgments

We thank the assistance of Amanda Backner, Jamie Heinzman, Sharon Clifford, and Shara Rice in the preparation of the manuscript.

References

1. Gallion, H., Christopherson, W. A., Coleman, R. L., et al. (2006) Progression-free interval in ovarian cancer and predictive value of an ex vivo chemoresponse assay. *Int J Gynecol Cancer* **16**, 194–201.
2. Ness, R. B., Wisniewski, S. R., Eng, H., and Christopherson, W. (2002) Cell viability assay for drug testing in ovarian cancer: in vitro kill versus clinical response. *Anticancer Res* **22**, 1145–1149.
3. Kornblith, P., Wells, A., Gabrin, M. J., et al. (2003) In vitro responses of ovarian cancer to platinums and taxanes. *Anticancer Res* **23**, 543–548.
4. Kornblith, P., Wells, A., Gabrin, M. J., et al. (2003) Breast cancer-response rates to chemotherapeutic agents studied in vitro. *Anticancer Res* **23**, 3405–3412.
5. Ochs, R. L., Fensterer, J., Ohori, N. P., et al. (2003) Evidence for the isolation, growth, and characterization of malignant cells in primary cultures of human tumors. *In Vitro Cell Dev Biol Anim* **39**, 63–70.
6. Ochs, R. L., Chattapadhyay, A., Bratton, R., et al. (2004) Phenotypic cell culture assay for predicting anticancer drug responses. *Preclinica* **2**, 205–212.
7. Samson, D. J., Seidenfeld, J., Ziegler, K., et al. (2004) Chemotherapy sensitivity and resistance assays: a systematic review. *J Clin Oncol* **22**, 3618–3630.
8. O'Meara, A. T., and Sevin, B.-U. (2001) Predictive value of the ATP chemosensitivity assay in epithelial ovarian cancer. *Gynecol Oncol* **83**, 334–342.
9. Kern, D. H., and Weisenthal, L. M. (1990) Highly specific prediction of antineoplastic drug resistance with an in vitro assay using suprapharmacologic drug exposures. *J Natl Cancer Inst* **82**, 582–588.
10. Chu, E., and DeVita, V. T. Jr. (2001) In vitro drug response assays, in *Cancer: Principles and Practice of Oncology* (DeVita, V. T. Jr., Hellman, S., and Rosenberg, S. A., eds.), Lippincott, Williams & Wilkins, Philadelphia, PA, pp. 302–304.

11. Fruehauf, J. P. (2002) In vitro assay-assisted treatment selection for women with breast or ovarian cancer. *Endocr. Relat. Cancer* **9**, 171–182.
12. Flowers, J. L., Ludeman, S. M., Gamcsik, M. P., et al. (2000) Evidence for a role of chloroethylaziridine in the cytotoxicity of cyclophosphamide. *Cancer Chemother Pharmacol* **45**, 335–344.

7

Correlation of Caspase Activity and In Vitro Chemo-Response in Epithelial Ovarian Cancer Cell Lines

Ayesha B. Alvero, Michele K. Montagna, and Gil Mor

Summary

The immediate assessment of response to therapy is most beneficial to ovarian cancer patients. This study shows the correlation of drug-induced caspase activation determined by western blot analysis and by Caspase-Glo™ assay. Our findings demonstrate that the use of the Caspase-Glo™ assay allows a simple, fast, and sensitive alternative for the evaluation of in vitro response to chemotherapy.

Key Words: Apoptosis; caspase-3 activity assay; ovarian cancer.

1. Introduction

Epithelial ovarian cancer (EOC) is the fourth leading cause of cancer-related deaths in women and is the most lethal of the gynecological malignances *(1)*. One of the major limitations in EOC treatment is the development of resistance to commonly used chemotherapeutic agents. This is confounded by the lack of practical means to predict clinical response.

It is now well documented that apoptosis or programmed cell death is the key mechanism by which chemotherapeutic agents exert their cytotoxicity *(2)*. Apoptosis is a cascade of intracellular factors known as caspases, which are highly specific proteases synthesized as zymogens and activated by cleavage *(3)*. Caspases can be divided into "initiators" or "effectors" of apoptosis. Initiator caspases, such as caspase-8 and caspase-9, mediate their oligomerization and autoactivation in response to upstream signals, whereas the effector caspases,

From: *Methods in Molecular Biology, vol. 414: Apoptosis and Cancer*
Edited by: G. Mor and A. B. Alvero © Humana Press Inc., Totowa, NJ

which include caspase-3, caspase-6, and caspase-7, cleave cellular substrates and precipitate apoptotic death.

Recently, we demonstrated a correlation between caspase-3 activity in EOC cells isolated from malignant ascites and the in vivo response to chemotherapy in patients with recurrent ovarian cancer *(4)*. We describe in this chapter the methods we used to correlate caspase activity and chemo-response in EOC.

2. Materials

2.1. Equipment

1. Cell lifter (Fischer Scientific, Pittsburg, PA, USA).
2. Tissue homogenizer.
3. 15- and 50-ml conical tubes (Falcon, BD Falcon, San Jose, CA, USA).
4. Cell-culture 96-well transparent, flat-bottom microplate (Falcon).
5. Disposable cuvettes (Promega, Madison, WI).
6. TD 20/20 lumonimeter.
7. SpectraMax M^2 microplate reader (Molecular Devices, Sunnydale, CA).

2.2. Reagents

1. Lysis buffer: 1× PBS + 1% NP40 + 0.1% sodium dodecyl sulfate.
2. Caspase-Glo™ reagents (Promega) (*see* **Note 1**).
3. Protease inhibitor.
4. Double-distilled water.

3. Methods

3.1. Cell Lysis

1. After the induction of apoptosis (*see* **Note 2**), scrape cells and collect in 15-ml tube.
2. Spin at 250 g for 10 min at 4°C.
3. Re-suspend pellet in 6 ml 1× PBS and spin again at 250 g for 10 min at 4°C to wash.
4. Remove as much PBS as possible without disrupting pellet and then add 100 μl lysis buffer with appropriate protease inhibitor (*see* **Note 3**).
5. Vortex and keep on ice for 20 min vortexing periodically.
6. Spin at 1100 g for 15 min at 4°C and collect supernatant. Keep lysates at –40°C until further use.

3.2. Tissue Lysis

1. Place 1 mg tissue in 5 ml ice-cold lysis buffer in a 50-ml conical tube.
2. Homogenize for 10 s maximum, three times at about 30-s interval (*see* **Note 4**). Homogenize in a circular motion pulling homogenizer up and down.

3. Wash homogenizer in cold water by turning on for 10 s, three times before homogenizing another tissue sample.
4. Transfer homogenized tissue into Eppendorf tubes and spin at 1100 g for 15 min at 4°C and collect supernatant. Keep lysates at –40°C until further use.

3.3. Caspase-Glo™ Assay for Lysates

1. Prepare Caspase-Glo™ reagents according to manufacturer's instruction.
2. Prepare 10 μg protein lysate in 50 μl total volume with double-distilled water in disposable cuvettes.
3. Incubate for 1 h at room temperature in the dark and then measure luminescence.

3.4. Caspase-Glo™ Assay for a 96-Well Plate

1. Plate cells on a 96-well plate (*see* **Note 5**) and induce apoptosis using selected method.
2. After the induction of apoptosis, directly add Caspase-Glo™ reagents (1:1) and incubate for 1 h in the dark at room temperature.
3. Measure luminescence using a plate reader.

4. Notes

1. Caspase-Glo™ –3/7, –8, and –9 reagents are separately available from Promega. However, other caspase activity assays may be used, but amount of reagents and incubation time may vary.
2. We typically induce apoptosis in EOC cells with 2 μM paclitaxel or 100 μg/ml carboplatin for 24 h.
3. We typically use protease inhibitor cocktail from Roche, Nutley, NJ, USA.
4. It is important to use ice-cold lysis buffer and to keep the 50-ml conical tube submerged on ice during homogenization to prevent rapid increase in temperature, which can denature proteins.
5. We typically plate 5000–6000 cells per well of EOC primary culture or 3000–4000 cells per well of established EOC cell lines. This yields approximately 70% confluent cultures the next day.

5. Conclusion

The availability of a sensitive and specific assay, which can provide immediate assessment to treatment response, is most beneficial to EOC patients. Our results show that the response of EOC cell lines to treatment can be evaluated in a couple of hours using the Caspase-Glo™ assay system. This is in contrast to performing western blot analysis, which can take up to 3 days before results can be obtained. The assays can be performed with whole cells in a 96-well plate or with cell and tissue protein extracts.

The clinical correlation of caspase activity and response to treatment is under investigation.

References

1. Schwartz, P.E. Current diagnosis and treatment modalities for ovarian cancer. *Cancer Treat Res* **107**, 99–118 (2002).
2. Kaufmann, S.H. Earnshaw, W.C. Induction of apoptosis by cancer chemotherapy. *Exp Cell Res* **256**, 42–9 (2000).
3. Cohen, G.M. Caspases: the executioners of apoptosis. *Biochem J* **326 (Pt 1)**, 1–16 (1997).
4. Flick, M.B. et al. Apoptosis-based evaluation of chemosensitivity in ovarian cancer patients. *J Soc Gynecol Investig* **11**, 252–9 (2004).

8

Assessing Expression of Apoptotic Markers Using Large Cohort Tissue Microarrays

Elah Pick, Mary M. McCarthy, and Harriet M. Kluger

Summary

Apoptotic markers include proteins from the intrinsic and extrinsic pathways. These cascades include both pro-apoptotic and anti-apoptotic elements. The expression levels of these elements can be assessed by immunohistochemistry (IHC) and can indicate general trends in pro- versus anti-apoptotic tendencies of the cells. IHC is particularly useful when studying large cohorts of paraffin-embedded specimens. Advances in tissue microarray (TMA) technology have facilitated evaluation of large cohorts of specimens, as cores from hundreds of patients can be represented on a single glass slide and stained in a uniform fashion. In this chapter, we discuss construction and staining methods of TMAs and present examples of assessment of apoptotic marker expression in malignant and benign cells using a novel method of automated, quantitative analysis of in situ protein expression.

Key Words: IHC; tissue microarrays; cohort size; immunofluorescence.

1. Introduction

One of the hallmarks of cancer is the ability to resist apoptosis in the face of external pro-death stimuli, such as chemotherapy, radiation therapy, and hypoxia *(1)*. Apoptotic pathways are highly complex. The two chief pathways known as the "direct" and the "indirect" pathways are described in detail in the literature *(2)*. The direct pathway, known also as the death receptor pathway, is triggered by external stimuli, which result in formation of a death-inducing signal complex that activates caspase-8. The indirect or the mitochondrial pathway is activated by extracellular stimuli as well as internal insults, such

From: *Methods in Molecular Biology, vol. 414: Apoptosis and Cancer*
Edited by: G. Mor and A. B. Alvero © Humana Press Inc., Totowa, NJ

as DNA damage. In this pathway, changes in the mitochondrial membrane potential occur through activation of pro-apoptotic members of the Bcl-2 family such as Bax, Bad, Bim, and Bid. Pro- and anti-apoptotic Bcl-2 family members meet on the mitochondrial surface where they compete in regulating the release of cytochrome c, which then associates with the Apaf-1/caspase-9 apoptosome. The two apoptotic pathways discussed above converge at the level of caspase-3 activation, which is antagonized by the IAP proteins. There are a number of pathways downstream of caspase-3, which results in cell degradation. Crosstalk between the death receptor and mitochondrial pathways is provided by Bid. The two pathways, however, are thought to predominantly act independently. Induction of apoptosis occurs not only by changes in expression levels of apoptotic mediators but also by changes in mitochondrial membrane potential and subcellular shifts of mediators such as Smac and Bax *(3)*. With the technology that we describe here, we can only assess changes in expression of mediators.

1.1. Tissue Microarray Construction

Immunohistochemistry (IHC) is a useful method for measuring expression levels of proteins in malignant and benign tumors. Tissue microarrays (TMAs) can represent large cohorts of tissue specimens arrayed in a single paraffin block. The modern TMA technology was developed as a response to the need for faster approaches to finding new biomarkers and is becoming a widely utilized tool in many pathology laboratories. It is also a practical method to screen antibodies or markers on the basis of patient outcome. Using this method, instead of staining hundreds of slides, an entire cohort of cases can be analyzed by staining slides from just one or two master blocks. Each spot on the array is similar to a conventional slide in that complete demographic and outcome information is maintained for each case so that rigorous statistical analysis can be done as rapidly as the arrays can be analyzed. The tissues are assembled by taking core needle "biopsies" of pre-existing paraffin-embedded tissues and re-embedding them in an arrayed "master" block, using techniques invented over 15 years ago by Wan and Furmanski *(4)* and an apparatus recently developed by Kononen et al. *(5)*.

TMAs have numerous advantages over traditional whole section staining; they essentially amplify the tissue available for analysis (around 1000-fold), as only thin, tiny cores of a specimen are used for staining each marker, rather than using an entire slice, as is done when staining whole sections. Other advantages include (i) the fact that all specimens are treated in a uniform, controlled fashion and (ii) only a very small (a few microliters) amount of antibody is required to analyze an entire cohort. A potential disadvantage of this technique is tissue and/or tumor heterogeneity. We and others have completed validation studies of TMAs *(6–8)*. Our work showed that, in breast cancer, equivalent results can be obtained in 95% of the cases by simply scoring two spots, suggesting a twofold redundant array is approximately representative of outcomes achieved using conventional histologic methods. This finding is complemented by work from other researchers. The most significant use of TMAs is in the evaluation of tissue biomarkers. This can be done by conventional pathologist-based analysis or by new methods discussed Section 1.4.

Storage of slides is critical. IHC of multiple tissue samples with standardized reagents and conditions was first reported in 1986 *(9)*. However, there was not

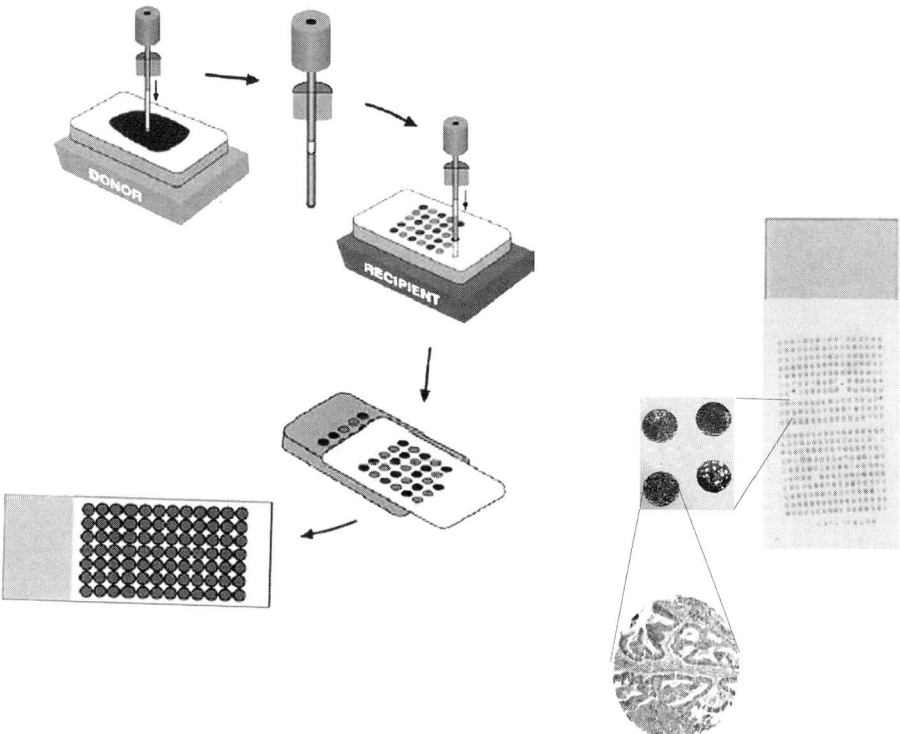

Fig. 1. A schema of construction of a tissue microarray. Cores from hundreds of donor blocks are placed in a recipient block, with careful database management. Slices are taken through the recipient block resulting in an array, as shown on the right. The inset on the right shows high magnification views of spots containing colon cancer.

enough data on the best method for preserving the sections. Once samples were cut and deparaffinized, the tissue was rapidly exposed to uncontrolled oxidation that could cause reduced antigenicity, resulting in experimental variability when performing stains at different times. Two techniques of paraffin-coating of slides and storage in a nitrogen desiccation chamber *(10)* have recently become standard for preserving the antigenicity of the tissue for a period of at least 3 months.

1.2. Cohort Size

A major advantage to TMAs is the ability to evaluate expression of a large number of markers in a large cohort of patients. One good example of the importance of cohort size is Her2/neu expression in melanoma. Owing to the success of trastuzumab in breast cancer, we evaluated the frequency of expression of Her2/neu. Previous studies, each including under 50 patients, reported huge variability in frequency of Her2/neu expression, ranging from 2.5 to 75%. We stained 600 cases for Her2/neu expression and found that only 5.2% expressed Her2/neu *(11)*.

1.3. Association Between Marker Expression and Clinical Outcome

Large cohort TMAs are particularly useful when associated clinical databases are available. Once expression levels have been measured, the scores can be entered into a database, allowing immediate assessment of association with clinical outcome parameters, such as tumor stage and survival. In previous studies, we assessed expression of pro- and anti-apoptotic markers and found associations with clinical outcome. Examples include expression of Bcl-2 in melanoma *(12)* and TRAIL receptor 1 (R1) and TRAIL R2 in breast cancer *(13)*. **Figure 2** shows the association between Bcl-2 expression and survival in melanoma.

1.4. Automated Analysis of TMAs

We have recently developed a fully quantitative method of analysis for TMAs. First, a series of images are collected by PM-1000, a custom microscope platform assembled in our laboratory from "off the shelf components," now commercially available (http://www.historx.com/). Each image set is analyzed by a system we call AQUA (Automated *QU*antitative *A*nalysis). AQUA allows measurements of protein expression within subcellular compartments that results in a number directly proportional to the number of molecules expressed per unit volume (the concentration). Unlike all previous methods for measurement of in situ protein expression (such as the CAS 200 or the ACIS

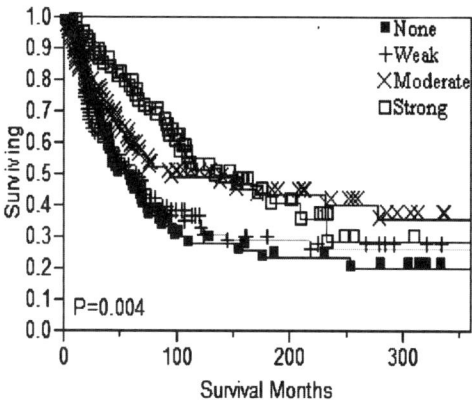

Fig. 2. Kaplan–Meier survival curves for Bcl-2 staining for 30 years of follow-up, with time given in months, showing the four quartiles of intensities of membranous Bcl-2 staining (logrank P = 0.004).

machine made by Chromavision, San Juan Capistrano, CA, USA), this method does not use feature extraction or "brown" stains. Instead, we use molecular methods to define subcellular compartments. We then quantify the amount of protein expressed within the compartment by co-localization. For example, to measure TRAIL R2 expression in breast cancer, the tissue is "masked" using cytokeratin in one channel to normalize to the area of tumor and to remove the stromal and other non-tumor material from the analysis. An image is then taken using DAPI to define a nuclear compartment. The pixels within the mask and within the DAPI-defined compartment are defined as nuclear. We measure the intensity of expression of TRAIL R2 using a third channel. The intensity of that subset of pixels divided by the number of pixels (to normalize the area from spot to spot) gives an AQUA score, which is directly proportional to the number of molecules of TRAIL R2 per unit area of tumor. This method including details of out-of-focus light subtraction imaging methods required to make this analysis work are described in detail *(14)* and illustrated schematically in **Fig. 3**. This technology has been licensed to HistoRx (New Haven, CT).

The strength of this method of analysis is illustrated in a series of recent papers that show results not discernable by conventional pathologist-based analysis. For example, examination of Her2/neu expression in a breast cancer cohort showed that high levels of expression were correlated to poor outcome. Surprisingly, a subset of the lowest expressers was found to do similarly poorly *(15)*. This finding has been previously described by a group using an ELISA

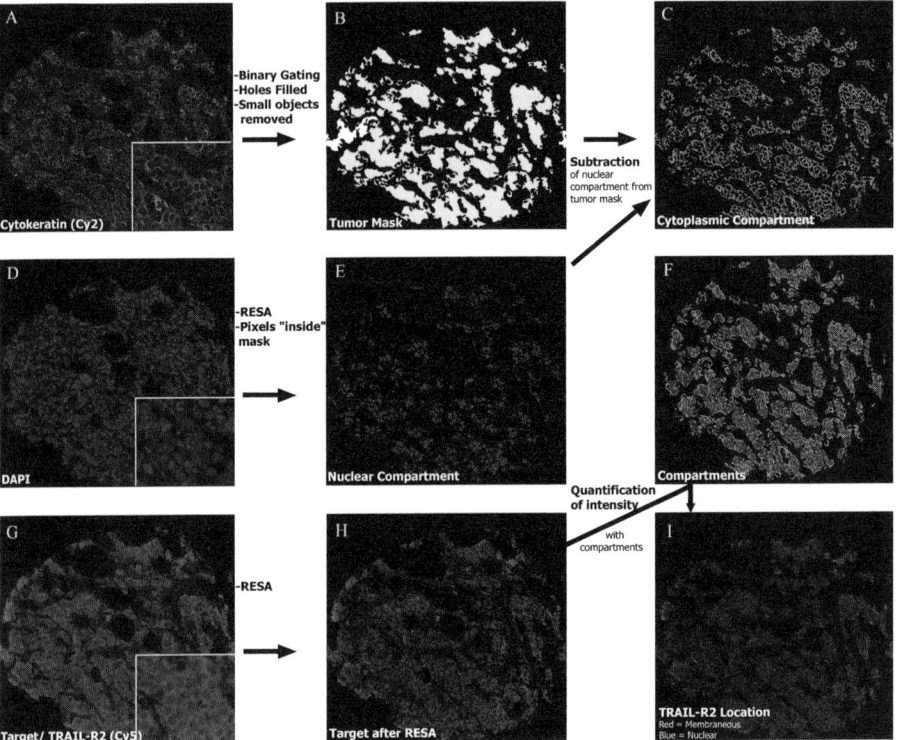

Fig. 3. Demonstration of Automated Quantitative Analysis (AQUA) used in quantitative measurement of TRAIL receptor 2 (R2) expression in breast cancer. Cytokeratin
is visualized using Cy2 to locate tumor (**A**). Holes are filled to create a tumor mask (**B**).
DAPI is used to locate nuclei (**D**). RESA (Rapid Exponential Subtraction Algorithm)
is used to eliminate background (**E**). Signal from the nuclei is subtracted from that
of the mask to create a membrane/cytoplasmic compartment (**C**). The nuclear and
membrane/cytoplasmic compartments are overlaid (**F**). Images are taken of the target
antigen (TRAIL R2) labeled using Cy5 tyramide (**G**). RESA is used to eliminate
background from the target (**H**), and TRAIL R2 within the compartments is measured
on a scale of 0–255 (**I**).

assay on extracted breast tissue (*16*). This relationship was not discernable by
conventional pathologists, as they are unable to distinguish "low" from "very
low." We have now identified a number of additional examples of expression
patterns associated with outcome that are not discernable with manual scoring
(*14,17*).

2. Materials

2.1. Equipment

1. TMA paraffin-coated slides are purchased from the Yale Tissue Microarray Core Facility (http://tissuearray.org/yale/tisarray.html) or other academic or commercial facilities (http://www.ihcworld.com/tissuearray.html). Prior to performing staining on a large-cohort TMA, we titer and validate our antibodies using "test arrays." Test array slides contain a limited numbers of tumors as well as a panel of cell lines and are used for optimizing the conditions and dilution of the primary antibody, assessing differential expression in cell lines that can be used to generate standard curves for the data, while avoiding excessive use of human tissue. After optimizing the conditions, full arrays with large cohorts of specimens can be used.
2. PM-1000 or the later version, PM-2000, a custom microscope platform with a camera and computer for moving the stage (http://www.historx.com).
3. Software for image analysis (AQUA) (http://www.historx.com).
4. Statistical software program such as JMP5, SAS, or SPSS.
5. Pressure cooker.

2.2. Reagents

Here, we use rabbit anti-TRAIL R2 in breast cancer as an example of a target marker for fluorescent staining. In this example, the mask is created by mouse anti-cytokeratin. The antibody against the target and antibody used to create the mask must be of a different species.

1. Xylene.
2. Ethanol.
3. Sodium citrate.
4. Citric acid.
5. 30% hydrogen peroxide.
6. Tween 20.
7. Tris buffer saline (TBS), pH 8.0.
8. TBS (pH 8.0) containing 0.005% Tween 20 (TBS/Tween).
9. TBS (pH 8.0) containing 0.3% bovine serum albumin (Sigma, St. Louis, MO, USA) (TBS/BSA).
10. TRAIL R2 polyclonal antibody (Oncogene Research, San Diego, CA).
11. Mammary epithelial marker: mouse anti-cytokeratin (DAKI Corporation, Carpinteria, CA, USA).
12. Envision goat anti-rabbit (DAKO Cytomation).
13. Goat anti-mouse immunoglobulin G conjugated to Alexa 488 (Molecular Probes Carlsbad, CA, USA).
14. Cy5 tyramide (Perkin-Elmer, Boston, MA).

15. Amplification diluent (Perkin-Elmer).
16. Prolong gold anti-fade reagent with DAPI (Invitrogen Carlsbad, CA, USA).

3. Methods

3.1. Fluorescent IHC

3.1.1. Before Starting

1. Label the slide in pencil withs the name, concentration of the antibody we are using, and the date we are performing our stain.
2. For titering, we recommend testing a minimum of four dilutions on four test arrays.

3.1.2. Deparaffinization

1. Deparaffinize the TMA slides with two changes of 100% xylene for 30 min each or until the paraffin is off.
2. Wash slides in two changes of 100% ethanol, 3 min each, and rehydrate the slides in tap water.

3.1.3. Antigen Retrieval

1. Antigen retrieval is performed by pressure cooking the slides for 15 min in 6.5 mM sodium citrate buffer (pH 6.0).

3.1.4. Block Peroxidases

1. To block endogenous peroxidases, incubate the slides in a mixture of methanol and 2.5% hydrogen peroxide for 30 min at room temperature.

3.1.5. Protein Blocking

1. Rinse the slides in tap water.
2. Dry the slides carefully around the array edge with a kimwipe.
3. To reduce nonspecific background staining (blocking), incubate the slides at room temperature for 30 min in TBS/BSA.

3.1.6. Target Identification

To select the primary antibodies for double staining, one has to stain the target (TRAIL R2) and the other one to visualize the tumor mask (cytokeratin).

1. Decant off the TBS/BSA and dry the edge with a kimwipe.
2. Incubate the slides at 4°C overnight in a humidity tray with the primary antibodies in rabbit polyclonal anti-TRAIL R2 immunoglobulin G at 1:350 (*see* **Subheading 3.1.9.**) and mouse monoclonal anti-cytokeratin immunoglobulin G at 1:200, in TBS/BSA. Add approximately 300 µl mixture containing the primary antibodies per slide.
3. On the second day, pour off the antibodies and wash the slides three times in TBS/Tween while gently shaking the slides in a box at room temperature.

3.1.7. Incubation

Dry the slides again around the array edge and add goat anti-mouse immunoglobulin G conjugated to Alexa 488 at a dilution of 1:100 to identify the mask (cytokeratin) and goat anti-rabbit horseradish peroxidase (Envision) to identify the target. The goat anti-horseradish peroxidase also serves as the diluent for Alexa 488.

1. Incubate in a dark humidity chamber at room temperature for 1 h.
2. Remove the secondary antibodies and wash the slides three times in TBS/Tween.
3. Dilute Cy5 directly conjugated to tyramide in amplification diluent, at a dilution of 1:50 for primary antibody identification.

3.1.8. Mounting the Slide

Rinse again and mount coverslips using prolong gold anti-fade reagent, which contains DAPI to visualize the nuclei.

1. Remove the bubbles and let it dry overnight in the dark.

3.1.9. Choice of Target Antibody Concentration

Optimal target antibody concentration is done either by direct visualization of the slides under a fluorescent microscope (choosing the concentration that demonstrates variability in staining intensity between tumors and normal tissues on the slide) or by collection of images Section 3.2–3.3 and choosing a concentration that yields differences in staining in the cell lines consistent with data on cell line expression of the target. The latter method is the preferred method if information is available on expression of the marker in cell lines embedded in the array.

3.2. Automated Image and Analysis

Using the PM-1000 or PM-2000 microscope platform, images are collected as described in detail *(18)*. Multiple monochromatic, high resolution (1024 × 1024 pixel, 0.5 μm) grayscale images are obtained for each histospot, using the 10× objective of the epifluorescence microscope. Two images (one in-focus and one out-of-focus) are taken of the compartment-specific tags (cytokeratin, TRAIL R2, and DAPI in this example).

3.3. Algorithmic Analysis

Using the AQUA software, a percentage of the out-of-focus image is subtracted from the in-focus image for each pixel, representing the signal-to-noise ratio of the image. An algorithm described as RESA (Rapid Exponential

Subtraction Algorithm) is used to subtract the out-of-focus information in a uniform fashion for the entire microarray. Subsequently, the PLACE (Pixel Locale Assignment for Compartmentalization of Expression) algorithm is used to assign each pixel in the image to a specific subcellular compartment, and the signal in each location is calculated. Pixels that cannot accurately be assigned to a compartment are discarded. The data are saved and subsequently expressed as the average signal intensity per unit of compartment area. For the nuclear and membrane/cytoplasmic compartments, the images are measured on a scale of 0–255 and expressed as target signal intensity relative to the compartment area.

3.4. Statistical Analysis

AQUA scores (or brown stain scores for traditional IHC) are imported into databases containing patient demographic, pathological, and clinical information. A number of software packages are available for data analyses such as those from SPSS (http://www.spss.com/) or SAS (http://www.sas.com/).

4. Notes

1. The fluorescent secondary antibodies are sensitive to light, and therefore staining and storage of the slides should be done in subdued light.
2. Careful design of test arrays is recommended, particularly in cases where there is abundant information on expression of targets in cell lines. Cell lines embedded in the arrays are used for normalization of scores obtained from different arrays as well as for generating standard curves for target expression.

References

1. Hanahan, D., and Weinberg, R.A. (2000) The hallmarks of cancer *Cell* **100**, 57–70.
2. Hengartner, M.O. (2000) The biochemistry of apoptosis. *Nature* **407**, 770–776.
3. Zhang, X.D., Borrow, J.M., Zhang, X.Y., Nguyen, T., and Hersey, P. (2003) Activation of ERK1/2 protects melanoma cells from TRAIL-induced apoptosis by inhibiting Smac/DIABLO release from mitochondria. *Oncogene* **22**, 2869–2881.
4. Wan, W.H., Fortuna, M.B., and Furmanski, P.A. (1987) rapid and efficient method for testing immunohistochemical reactivity of monoclonal antibodies against multiple tissue samples simultaneously. *J Immunol Methods* **103**, 121–129.
5. Kononen, J., Bubendorf, L., Kallioniemi, A., Barlund, M., Schraml, P., Leighton, S., Torhorst, J., Mihatsch, M.J., Sauter, G., and Kallioniemi, O.P. (1998) Tissue microarrays for high-throughput molecular profiling of tumor specimens. *Nat Med* **4**, 844–847.
6. Bova, G.S., Parmigiani, G., Epstein, J.I., Wheeler, T., Mucci, N.R., and Rubin, M.A. (2001) Web-based tissue microarray image data analysis: initial validation testing through prostate cancer Gleason grading. *Hum Pathol* **32**, 417–427.

7. Camp, R.L., Charette, L.A., and Rimm, D.L. (2000) Validation of tissue microarray technology in breast carcinoma. *Lab Invest* **80**, 1943–1949.

8. Hoos, A., Urist, M.J., Stojadinovic, A., Mastorides, S., Dudas, M.E., Leung, D.H., Kuo, D., Brennan, M.F., Lewis, J.J., and Cordon-Cardo, C. (2001) Validation of tissue microarrays for immunohistochemical profiling of cancer specimens using the example of human fibroblastic tumors. *Am J Pathol* **158**, 1245–1251.

9. Giltnane, J.M., and Rimm, D.L. (2004) Technology insight: identification of biomarkers with tissue microarray technology. *Nat Clin Pract Oncol* **1**, 104–111.

10. DiVito, K.A., Charette, L.A., Rimm, D.L., and Camp, R.L. (2004) Long-term preservation of antigenicity on tissue microarrays. *Lab Invest* **84**, 1071–1078.

11. Kluger, H.M., DiVito, K., Berger, A.J., Halaban, R., Ariyan, S., Camp, R.L., and Rimm, D.L. (2004) Her2/neu is not a commonly expressed therapeutic target in melanoma. *Melanoma Res* **14**, 207–210.

12. Divito, K.A., Berger, A.J., Camp, R.L., Dolled-Filhart, M., Rimm, D.L., and Kluger, H.M. (2004) Automated quantitative analysis of tissue microarrays reveals an association between high Bcl-2 expression and improved outcome in melanoma. *Cancer Res.* **64**, 8773–8777.

13. McCarthy, M.M., Sznol, M., DiVito, K.A., Camp, R.L., Rimm, D.L., and Kluger, H.M. (2005) Evaluating the expression and prognostic value of TRAIL-R1 and TRAIL-R2 in breast cancer. *Clin Cancer Res* **11**, 5188–5194.

14. Camp, R.L., Chung, G.G., and Rimm, D.L. (2002) Automated subcellular localization and quantification of protein expression in tissue microarrays. *Nat Med* **8**, 1323–1327.

15. Camp, R.L., Dolled-Filhart, M., King, B.L., and Rimm, D.L. (2003) Quantitative analysis of breast cancer tissue microarrays shows that both high and normal levels of HER2 expression are associated with poor outcome. *Cancer Res* **63**, 1445–1448.

16. Koscielny, S., Terrier, P., Spielmann, M., and Delarue, J.C. (1998) Prognostic importance of low c-erbB2 expression in breast tumors. *J Natl Cancer Inst* **90**, 712.

17. Rubin, M.A., Zerkowski, M.P., Camp, R.L., Kuefer, R., Hofer, M.D., Chinnaiyan, A.M., and Rimm, D.L. (2004) Quantitative determination of expression of the prostate cancer protein alpha-methylacyl-CoA racemase using automated quantitative analysis (AQUA): a novel paradigm for automated and continuous biomarker measurements. *Am J Pathol* **164**, 831–840.

18. Camp, R.L., Chung, G.G., and Rimm, D.L. (2002) Automated subcellular localization and quantification of protein expression in tissue microarrays. *Nat Med* **11**, 1323–1327.

9

Mitochondria Potential, Bax "Activation," and Programmed Cell Death

C. Michael Knudson and Nicholas M. Brown

Summary

Since the discovery of the key role of cytochrome C in the activation of caspase 9, intense interest has focused on the role of mitochondria in apoptosis/programmed cell death. Mitochondria undergo two major alterations during apoptosis. The first is the permeabilization of the outer mitochondrial membrane. This event is tightly regulated by members of the Bcl-2 family and involves the conformational change of pro-apoptotic family members such as Bax. Second, the electrochemical gradient that is normally present across the inner mitochondrial membrane is lost (membrane depolarization). This event is sometimes mediated by the permeability transition pore (PTP). The order in which these events occur and whether one causes the other has been hotly debated in the literature. Nonetheless, the majority of reports suggest that mitochondria outer membrane permeabilization (MOMP) precedes membrane depolarization. In this chapter, methods that examine membrane depolarization and the conformational change in Bax are described.

Key Words: Apoptosis, Bax, Mitochondria, Cytochrome C.

1. Introduction

Since the term apoptosis was originally proposed, the role of mitochondria in this process has been debated *(1)*. Initial reports suggested that mitochondria structure was preserved during apoptosis in contrast to disruption and swelling of mitochondria during necrotic cell death *(2)*. This dogma was largely retained until the revelation that the mitochondrial protein cytochrome C was involved in the activation of caspase 9, a critical apical caspase in the caspase cascade *(3,4)*. These findings, and many since, have confirmed that mitochondria are

From: *Methods in Molecular Biology, vol. 414: Apoptosis and Cancer*
Edited by: G. Mor and A. B. Alvero © Humana Press Inc., Totowa, NJ

intimately linked to the regulation of apoptotic cell death. As cytochrome C exists in the intermembrane space, much effort has focused on how the outer mitochondria membrane becomes permeable to cytochrome C and allows its release from the organelle. Two competing, although not mutually exclusive, hypotheses have been proposed.

One proposal centers around the role of membrane depolarization in apoptosis. In this model, the permeability transition pore (PTP) results in rapid depolarization of the inner mitochondria membrane. The outer mitochondria membrane then becomes permeable to cytochrome C by direct mechanical (swelling) or alternative pathways. The second proposal suggests that mitochondria outer membrane permeabilization (MOMP) precedes inner membrane depolarization and is directly regulated by the Bcl-2 family. In this model, the multi-domain pro-apoptotic members of the Bcl-2 family (Bax and Bak) are critical for this event. The "BH3 only," pro-apoptotic members of the Bcl-2 family and the anti-apoptotic family members are involved in a "life and death" struggle to either activate or prevent the activation of Bax and/or Bak *(5)*. The critical (and redundant) role of Bax and Bak in this process has been convincingly demonstrated by knockout studies in which loss of both Bax and Bak make cells very resistant to release of cytochrome C and apoptosis *(6–8)*. In this model, Bax and Bak are present in the cell but kept in an "inactive" state by anti-apoptotic Bcl-2 family members. Following an apoptotic signal, BH3 family members are either transcriptionally up-regulated or activated by post-translational modifications that then result, either directly or indirectly, in the "activation" of Bax and/or Bak.

Richard Youle and colleagues *(9–13)* were the first to report that Bax subcellular distribution was altered during apoptosis. In these studies, they demonstrated that Bax exists in both the cytoplasm/soluble fraction and the membrane-associated fraction. During apoptosis, although the total levels of Bax are generally not changed, the subcellular distribution of Bax is changed dramatically from a soluble/cytoplasmic location to a membrane-associated fraction. Furthermore, this change in distribution is associated with a conformational change in Bax that exposes an amino-terminal epitope that can be detected with antibodies in this domain. These studies demonstrate that this conformation change is associated with the redistribution to mitochondria and the release of cytochrome C. Detection of the conformation change in Bax with amino-terminal antibodies has thus been used as a "surrogate" marker of outer mitochondria membrane permeabilization.

As mentioned, membrane depolarization is also observed during apoptosis and other types of cell death. Fortunately, several chemicals are available to easily detect mitochondria membrane depolarization. In general, these are relatively inexpensive, fluorescent compounds that are concentrated in mitochondria with an intact electrochemical gradient. During apoptosis or other types of cell death, this electrochemical gradient is lost and the fluorescence staining by these compounds is decreased in apoptotic cells. These assays, although not specific to apoptosis, offer an inexpensive, high-throughput method to assess levels of cell death. Multiple reagents have been used to measure mitochondria membrane potential *(14)*. In this review, we will focus on three reagents that have frequently been used in the literature. In addition, methods to simultaneously measure annexin V staining and mitochondria membrane potential function will also be described. Finally, we will describe methods to detect the conformation change in Bax that is thought to precede the release of cytochrome C and drop in mitochondrial membrane potential.

2. Materials

2.1. Equipment

1. Routine tissue-culture equipment.
2. Cells of interest.
3. Tubes for flow cytometer (cat. no. 352052, BD Falcon, Franklin Lakes, NJ, USA).
4. Flow cytometer with excitation laser of 488 nM (Becton Dickinson FacsScan, Franklin Lakes, NJ, USA). For double labeling, a dual laser instrument is preferred (Becton Dickinson FacsCalibur)

2.2. Reagents

1. 3,3′-dihexyloxacarbocyanine iodide (DiOC$_6$) from (cat. no. D273, Invitrogen Carlsbad, CA, USA).
2. Tetramethylrhodamine, methyl ester-perchlorate (TMRM) (cat. no. T668, Invitrogen).
3. 5,5′,6,6′-tetrachloro-1,1′,3,3′-tetraethylbenzimidazolylcarbocyanine iodide (JC-1) (cat. no. T-3168, Invitrogen).
4. Carbonyl cyanide *m*-chlorophenylhydrazone (CCCP) (cat. no. C2759, Sigma, St. Louis, MO, USA).
5. Annexin V-Cy5 and buffer (cat. no. K103-100, Biovision, Martin View, CA, USA).
6. Digitonin (cat. no. D-141, Sigma) (*see* **Note 1**).
7. Anti-FC monoclonal antibody—2.4G2 (Becton Dickinson).
8. Anti-Bax antibody—clone 6A7 (cat. no. 14-6997, eBioscience, San Pieso, CA, USA).
9. Isotype control antibody—IgG1-kappa (cat. no. 14-4714, eBioscience).
10. FITC-conjugated goat anti-mouse IgG antibody (cat. no. D554001, Becton Dickinson).

11. RNase (Fermentas, Glen Burnie, MD, USA).
12. Propidium iodide (PI) (Sigma).
13. Normal rat serum (available from many vendors including Sigma).
14. Paraformaldehyde (Fisher Scientific, Hampton, NH, USA) (prepare 8% stock in water—*see* **Note 2**).

3. Methods

Cytokine deprivation is a classic signal for programmed cell death. FL5.12 cells are an immortalized pre-B cell line derived from normal fetal liver that are dependent on interleukin-3 (IL-3) for their growth and survival *(15–17)*. Upon IL-3 removal/withdrawal, the cells undergo programmed cell death that can be blocked by expression of Bcl-2 *(18)* or delayed by activation of AKT *(15)*. Cell death occurs progressively after approximately 12–18 h with the majority of cells dead by 48 h.

3.1. FL5.12 Cell Culture and IL-3 Deprivation

1. FL5.12 cells (grown at 37°C, 5% CO2) are passaged (non-adherent so trypsin not necessary) every other day (~1:10) in FL5.12 media to maintain cell concentration less than 1.2×10^6 cells/ml.
2. Cells can be directly studied in this media for experiments not requiring IL-3 deprivation (e.g., CCCP treatment).
3. For IL-3 deprivation, FL5.12 cells are washed by centrifugation ($1200 \times g$ for 7 min) and resuspension with PBS. Repeat this step.
4. After second centrifugation, the cells are resuspended in FL5 media without IL-3 free media at 200,000–800,000 cells/ml.
5. Cells are returned to the incubator and cultured until the time of the experiment (generally 24–48 h).

3.2. DiOC₆ Staining

DiOC$_6$ is a dye that detects the proton gradient across the mitochondrial inner membrane and effectively measures mitochondrial membrane potential. Results with the DiOC$_6$ assay correlate very closely with other apoptotic assays such as trypan blue exclusion, PI exclusion visualized with a flow cytometer, and Guava viability analysis (data not shown). DiOC$_6$ staining by flow cytometry is an inexpensive, high-throughput method to assess cell death.

1. Prepare a stock solution of DiOC$_6$ in 100% EtOH at 1 mM (stable if stored at −20°C).
2. Then dilute the stock solution to 5 μM in media.
3. Add FL5.12 cells in media to a flow cytometry tube at 0.2–1.0 million cells/ml in a total volume of at least 0.3 ml.

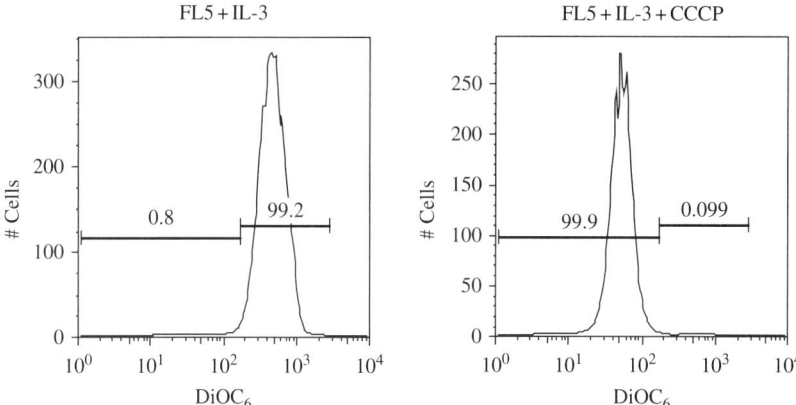

Fig. 1. 3,3′-dihexyloxacarbocyanine iodide (DiOC$_6$) staining of FL5.12 cells with (right) and without (left) 500 μM carbonyl cyanide *m*-chlorophenylhydrazone (CCCP) for 10 min prior to staining with 50 nM DiOC$_6$. Cells were then analyzed with a FacsCalibur as described in **Subheading 3**. The percent of total cells in the DiOC$_6$ low versus high gate is indicated above the marker.

4. Optional: as a control, add 100–500 μM CCCP to the cells and incubate for 10 min prior to DiOC$_6$ addition. CCCP cells should indicate reduced DiOC$_6$ staining relative to untreated viable cells (*see* **Fig. 1**).
5. Add DiOC$_6$ to a final concentration of 50 nM (*see* **Note 3**).
6. Incubate cells with DiOC$_6$ at 37°C for 10 min.
7. Acquire cells utilizing an appropriate flow cytometer [excitation wavelength = 488 nM; detection in the FL1 channel (530/30 nM bandpass filter)].
8. Perform data analysis using either Cell Quest software (Becton Dickinson) or FlowJo software.
9. Distinguish apoptotic cells that demonstrate a DiOC$_6$ low population from the viable cells (*see* **Fig. 2**).

3.3. TMRM Staining

TMRM is the methyl ester of tetramethylrhodamine and has been used as a rapid method to measure mitochondria membrane potential *(19)*. Both TMRM and TMRE (ethyl ester) are lipophilic cations that are reported to accumulate in mitochondria in proportion to membrane potential *(19)*. The measurement of TMRM low cells corresponds very well with cell death as measured by trypan blue exclusion, PI exclusion viewed with a flow cytometer, and Guava viability

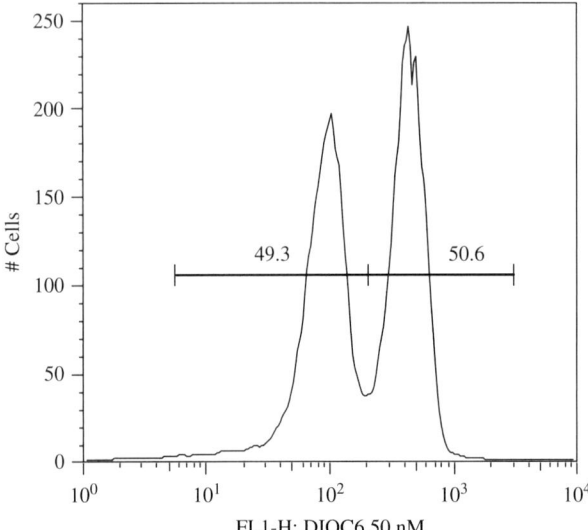

Fig. 2. 3,3´-dihexyloxacarbocyanine iodide (DiOC$_6$) staining of FL5.12 cells after 24 h of interleukin-3 (IL-3) deprivation. Cells were stained and analyzed as in **Fig. 1**. The percent of total cells in the DiOC$_6$ low (apoptotic) versus high (viable) gate is indicated above the marker.

analysis (data not shown). Thus, TMRM staining provides another method to easily and inexpensively measure cell death.

1. Prepare a stock solution of TMRM in DMSO at 1 mM (stable if stored in the dark at –20°C).
2. Then dilute the stock solution to 50 μM in media.
3. Add FL5.12 cells in media to a flow cytometry tube at 0.2–1.0 million cells/ml in a total volume of at least 0.3 ml.
4. Optional: as a control, add 100–500 μM CCCP to the cells and incubate for 10 min prior to TMRM addition (e.g., *see* **Fig. 3** of expected staining with CCCP).
5. Add TMRM to a final concentration of 2 μM.
6. Incubate cells with TMRM at 37°C for 10 min.
7. Acquire cells utilizing an appropriate flow cytometer [excitation wavelength = 488 nM; detection in the FL2 channel (585/42 nM bandpass filter)].
8. Perform data analysis using either Cell Quest software (Becton Dickinson) or FlowJo software.
9. Distinguish apoptotic cells that demonstrate a low TMRM fluorescence from the viable cells (*see* **Fig. 3**).

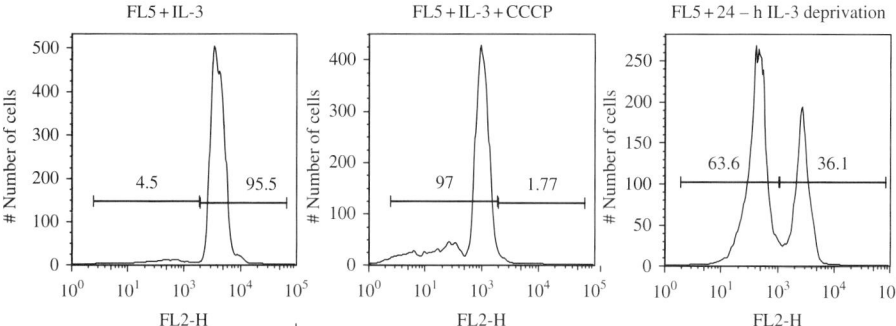

Fig. 3. Tetramethylrhodamine, methyl ester-perchlorate (TMRM) staining of FL5.12 cells with 500 uM carbonyl cyanide *m*-chlorophenylhydrazone (CCCP) (middle) for 10 min prior to staining, without CCCP (left), and with 24 h of interleukin-3 (IL-3) deprivation (right). Cells were stained with 50 nM TMRM for 10 min at 37°C and analyzed with a FacsCalibur. The percent of total cells in the TMRM low (dead) versus high (live) gate is indicated above the marker.

3.4. JC-1 Staining

JC-1 is a lipophilic cation cyanine molecule that has also been used to measure mitochondrial membrane potential *(20)*. However, unlike $DiOC_6$ and TMRM, JC-1 fluorescence is examined by looking at the relative fluorescence in two channels simultaneously (FL1 and FL2). This is due to the concentration-dependent fluorescence properties of this dye. At low concentration, the dye is present in monomeric form, and the fluorescence maxima can be detected in the FL1 channel. In contrast, at higher concentrations, JC-1 aggregates form that have a fluorescence maxima at a higher wavelength that is detected in the FL2 channel *(20)*. The accumulation or concentration of the dye is dependent on the mitochondria membrane potential. These properties make JC-1 a sensitive method to detect changes in membrane potential.

1. Prepare a stock solution of JC-1 in DMSO at 2.5 mg/ml (stable if stored at −20°C).
2. Just prior to the experiment, prepare a 10× stock solution of JC-1 (100 µg/ml in media).
3. Add FL5.12 cells in media to a flow cytometry tube at 0.2–1.0 million cells/ml in a total volume of at least 0.3 ml.
4. Optional: as a control, add 100–500 µM CCCP to the cells and incubate for 10 min prior to JC-1 addition (e.g., *see* **Fig. 4** of expected staining with CCCP).
5. Add JC-1 to a final concentration of 10 µg/ml.
6. Incubate cells with JC-1 at 37°C for 10 min.

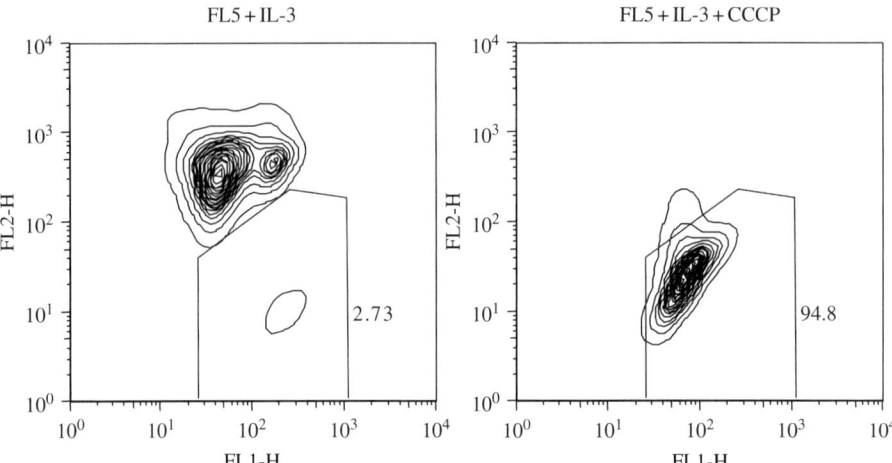

Fig. 4. JC-1 staining of FL5.12 cells with (right) and without (left) 500 uM carbonyl cyanide *m*-chlorophenylhydrazone (CCCP) for 10 min prior to staining with 10 ug/ml JC-1. Cells were then analyzed with a FacsCalibur as described in **Subheading 3**. The percent of total cells in the "FL-2 low" gate (apoptotic) is indicated next to the marker.

7. Acquire cells utilizing an appropriate flow cytometer [excitation wavelength = 488 nM; detection in the FL1 channel (530/30 nM bandpass filter) and FL2 channel (585/42 nM bandpass filter)].
8. Adjust the compensation so that the optimal separation of the viable and dead cells is obtained. The emission spectra of JC-1 monomers and JC-1 aggregates overlap. IL-3-deprived FL5.12 cells containing both viable and dead populations are shown with and without compensation (*see* **Fig. 5**). Optimal compensation settings will vary from cell type to cell type and even from experiment to experiment (*see* **Note 4**).
9. Perform data analysis using either Cell Quest software (Becton Dickinson) or FlowJo software.
10. Notice apoptotic cells generally demonstrate low-intensity staining for FL2 (aggregates) and high-intensity staining for FL1 (monomers). We have found that the loss of FL2 staining is a more reliable marker of apoptosis than the increase in FL1 staining (*see* **Fig. 5**, left panel).

3.5. Annexin V Staining

It is possible to stain for mitochondrial membrane potential using any of the previously described dyes while simultaneously staining with a second apoptotic indicator. Here, we describe staining the cells with annexin V, which

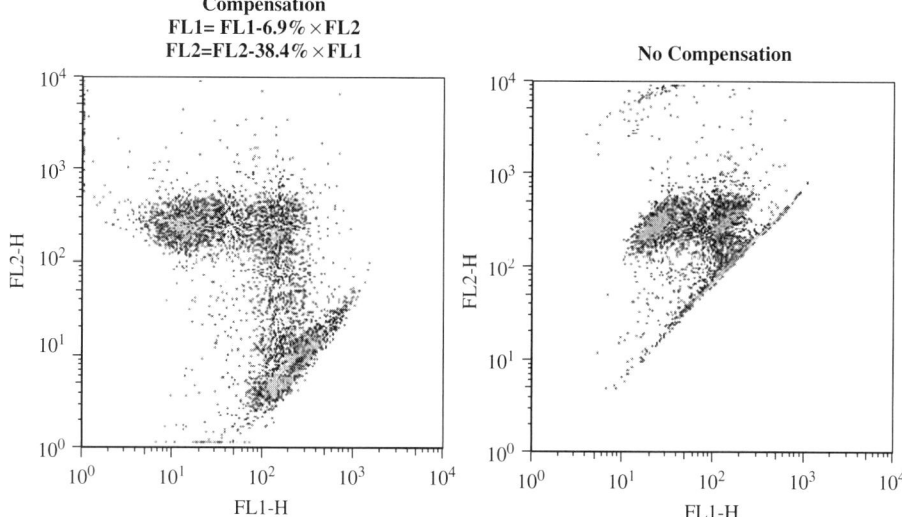

Fig. 5. FL5.12 cells were deprived of interleukin-3 (IL-3) for 24 h, then stained with 10 ug/ml JC-1 for 10 min at 37°C. Cells were then acquired with (left) and without (right) compensation using a FacsCalibur flow cytometer.

detects the exposure of phosphatidylserine on the cell surface, an early marker of apoptosis *(21)*. This allows for the examination of membrane potential for annexin V-positive versus annexin V-negative populations. Studies such as these provide insight into the order of events during apoptosis.

1. Stain for mitochondrial potential as described in Sections 3.2, 3.3 or 3.4. Staining should be performed in 1.5-ml Eppendorf tubes because the cells require centrifugation for annexin V staining.
2. Centrifuge cells in a microcentrifuge for 7 min at 1200 g and aspirate media.
3. Resuspend cells in 0.25 ml annexin V buffer with 1:1000 annexin V-Cy5 and transfer to flow tubes (*see* **Note 5**).
4. Incubate cells in the dark for 5 min at room temperature. They are then placed on ice prior to data acquisition.
5. Acquire cells utilizing an appropriate flow cytometer [excitation wavelength = 635 nM; detection in the FL4 channel (661/16 nM bandpass filter)]. The acquisition of mitochondrial potential as described in Section 3.2, 3.3 or 3.4 can be done simultaneously utilizing the other excitation laser and detection channels. As a second laser is utilized to detect Cy5, no compensation is necessary between the mitochondrial potential dyes and the annexin V channel.

6. Perform data analysis using either Cell Quest software (Becton Dickinson) or FlowJo software.
7. Simultaneous staining with DiOC$_6$ and Annexin V in FL5.12 following IL-3 deprivation indicates that nearly 100% of the Annexin V positive cells have low DiOC$_6$ staining while a significant fraction (~20% of Annexin V negative cells stain brightly for DiOC$_6$) (**Fig. 6**).

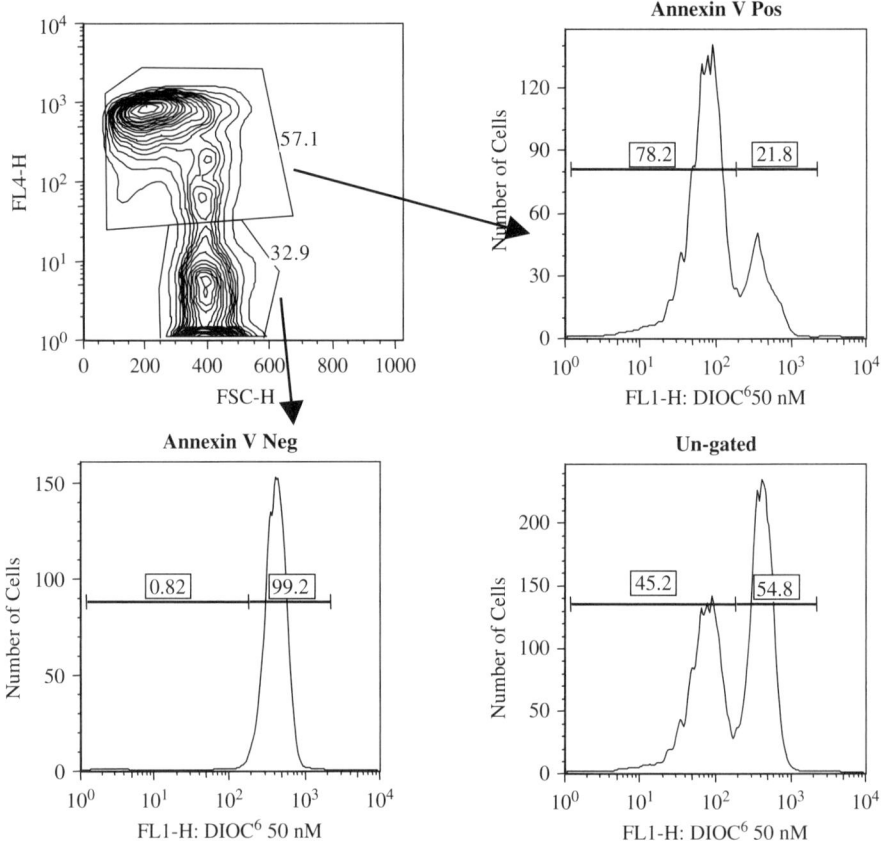

Fig. 6. Cells were stained with 3,3′-dihexyloxacarbocyanine iodide (DiOC$_6$) and annexin V-Cy5 following 24-h interleukin-3 (IL-3) deprivation and analyzed with a FacsCalibur as described in **Subheading 3**. The upper left plot shows percentages of annexin V low (viable) and high (apoptotic) cells. The remaining plots show DiOC$_6$ staining of the annexin V-negative population (bottom left), the annexin V-positive population (upper right), and ungated population (bottom right).

3.6. "Activated" Bax Staining by Flow Cytometry

This section describes staining for the N-terminal epitope of Bax that is exposed following Bax mitochondria localization *(16–24)*.

1. Pellet $1–2 \times 10^6$ FL5.12 cells by centrifugation ($1000 \times g$ for 7 min) and resuspend in a small volume (typically 100 µl) of PBS.
2. Add an equal volume of PBS with 0.5% paraformaldehyde and incubate for 5 min at room temperature to fix the cells.
3. Wash cells with PBS and resuspend in 100 µl blocking buffer [PBS with 0.01% digitonin, 5 µg/ml anti-FC (monoclonal Ab-2.4G2), and 10 % normal rat serum].
4. Incubate at room temperature for 5–10 min.
5. Add anti-Bax antibody (clone 6A7) or isotype control (IgG1-kappa) at 2.5 µg/sample and incubate at room temperature for 30 min.
6. Wash twice with PBS/0.01% digitonin.
7. Following the second wash, resuspend the pellet cells in 100 µl buffer (PBS/0.01% digitonin) with 1 µg FITC-conjugated goat anti-mouse IgG antibody.
8. Incubate for 30 min at 4°C protected from light.
9. Wash twice with PBS/0.01% digitonin.
10. After the second wash, resuspend the cells with 300–400 µl DNA staining solution (PBS with 5 µg/ml PI and 50 µg/ml RNase A).

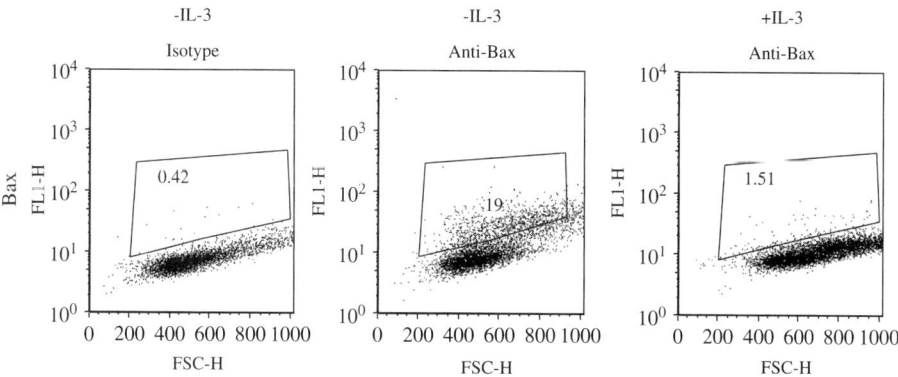

Fig. 7. Staining for activated Bax. FL5.12 cells were deprived of interleukin-3 (IL-3) for 24 h or maintained in IL-3 as indicated. Cells were fixed and stained with a monoclonal antibody that detects that "active" conformation of Bax or an isotype control antibody. All cells were then stained with a GAM-FITC. Cells were counterstained with propidium iodide (PI) to detect DNA content, and the FITC staining (FL1) of the cells that were viable by DNA content is shown. The percentage of cells that shown significant staining with FITC is indicated in the region.

11. Incubate in the dark at room temperature for 20 min to allow RNase A to work.
12. Acquire cells utilizing an appropriate flow cytometer [excitation wavelength = 488 nM; detection in the FL1 channel (530/30 nM bandpass filter) and FL2 channel (585/42 nM bandpass filter)]. Acquisition should be performed with doublet discrimination turned on for the FL2 channel. In these studies, FL2 staining can vary from sample to sample, so the FL2 gain is adjusted so that the diploid cells run at the same location for all the samples (we center them around 400). During analysis of FITC staining, doublets and subdiploid cells are excluded utilizing gates within the FL2-A versus FL2-W plots.
13. Perform data analysis using either Cell Quest software (Becton Dickinson) or FlowJo software (*see* **Note 5**).
14. Talk about results here (*see* **Fig. 7**).

4. Notes

1. The quality of the digitonin is critical; highly purified digitonin is required such as that described here from Sigma. Lower grade digitonin will remain insoluble and not permeabilize cells at these concentrations.
2. An 8% aqueous stock is prepared by adding 8 g/10 ml water in a 15-ml conical tube. Two drops of 10 N NaOH are added, and the tube is placed into boiling water until the paraformaldehyde dissolves. This generally takes 5–10 min. This 8% solution can then be stored at 4°C and should be prepared fresh every month.
3. The optimal concentration of each dye can be determined by looking at the difference in fluorescence in the presence and absence of CCCP. For CCCP treatment, cells should be treated with 50–500 μM drug for at least 10 min at 37°C *before* the cells are stained with the respective dye. The concentration of dye that gives that largest relative shift with CCCP should be chosen for subsequent experiments.
4. This is not compensation in the "traditional" sense because there are no positive controls with only FL-1 or FL-2 emission. This makes compensation much more subjective for JC-1 than it is for traditional staining protocols. A qualitative analysis of uncompensated data demonstrates that there is significant "bleeding" of the monomeric emission into the FL-2 channel. JC-1 aggregate emission also shows up to a lesser degree in the FL-1 channel. We have found that compensation settings of FL1 = FL1 – 6.9 × FL2 and FL2 = FL2 – 38.4 × FL1 generally work well for FL5.12 cells under the conditions described. However, this may vary dramatically for other systems and should be optimized for each experiment.
5. Optimal concentration of annexin V-Cy5 should be determined for each cell type. FL5.12 cells stain well brightly with annexin and require much less reagent than the manufacturer recommends.

References

1. Kerr, J. F., A. H. Wyllie, and A. R. Currie. 1972. Apoptosis: a basic biological phenomenon with wide-ranging implications in tissue kinetics. *British Journal of Cancer 26:239–257.*

2. Wyllie, A. H., J. F. Kerr, and A. R. Currie. 1980. Cell death: the significance of apoptosis. *International Review of Cytology 68:251–306.*

3. Liu, X. S., C. N. Kim, J. Yang, R. Jemmerson, and X. D. Wang. 1996. Induction of apoptotic program in cell-free extracts - requirement for datp and cytochrome c. *Cell 86:147–157.*

4. Li, P., D. Nijhawan, I. Budihardjo, S. M. Srinivasula, M. Ahmad, E. S. Alnemri, and X. Wang. 1997. Cytochrome c and dATP-dependent formation of Apaf-1/Caspase-9 complex initiates an apoptotic protease cascade. *Cell 91:479–489.*

5. Kuwana, T., and D. D. Newmeyer. 2003. Bcl-2-family proteins and the role of mitochondria in apoptosis. *Current Opinion in Cell Biology 15:691–699.*

6. Cheng, E., M. C. Wei, S. Weiler, R. A. Flavell, T. W. Mak, T. Lindsten, and S. J. Korsmeyer. 2001. BCL-2, BCL-X-L sequester BH3 domain-only molecules preventing BAX- and BAK-mediated mitochondrial apoptosis. *Molecular Cell 8:705–711.*

7. Lindsten, T., A. J. Ross, A. King, W. X. Zong, J. C. Rathmell, H. A. Shiels, E. Ulrich, K. G. Waymire, P. Mahar, K. Frauwirth, Y. F. Chen, M. Wei, V. M. Eng, D. M. Adelman, M. C. Simon, A. Ma, J. A. Golden, G. Evan, S. J. Korsmeyer, G. R. MacGregor, and C. B. Thompson. 2000. The combined functions of proapoptotic Bcl-2 family members Bak and Bax are essential for normal development of multiple tissues. *Molecular Cell 6:1389–1399.*

8. Wei, M. C., W. X. Zong, E. H. Y. Cheng, T. Lindsten, V. Panoutsakopoulou, A. J. Ross, K. A. Roth, G. R. MacCregor, C. B. Thompson, and S. J. Korsmeyer. 2001. Proapoptotic BAX and BAK: A requisite gateway to mitochondrial dysfunction and death. *Science 292:727–730.*

9. Wolter, K. G., Y. T. Hsu, C. L. Smith, A. Nechushtan, X. G. Xi, and R. J. Youle. 1997. Movement of Bax from the cytosol to mitochondria during apoptosis. *The Journal of Cell Biology 139:1281–1292.*

10. Hsu, Y. T., and R. J. Youle. 1997. Nonionic detergents induce dimerization among members of the Bcl-2 family. *The Journal of Biological Chemistry 272: 13829–13834.*

11. Hsu, Y. T., K. G. Wolter, and R. J. Youle. 1997. Cytosol-to-membrane redistribution of Bax and Bcl-X(L) during apoptosis. *Proceedings of the National Academy of Sciences of the United States of America 94:3668–3672.*

12. Hsu, Y. T., and R. J. Youle. 1998. Bax in murine thymus is a soluble monomeric protein that displays differential detergent-induced conformations. *The Journal of Biological Chemistry 273:10777–10783.*

13. Nechushtan, A., C. L. Smith, Y. T. Hsu, and R. J. Youle. 1999. Conformation of the Bax C-terminus regulates subcellular location and cell death. *The EMBO Journal 18:2330–2341.*

14. Zamzami, N., C. Maisse, D. Metivier, and G. Kroemer. 2001. Measurement of membrane permeability and permeability transition of mitochondria. *Methods in Cell Biology 65:147.*

15. Plas, D. R., S. Talapatra, A. L. Edinger, J. C. Rathmell, and C. B. Thompson. 2001. Akt and Bcl-x(L) promote growth factor-independent survival through distinct effects on mitochondrial physiology. *The Journal of Biological Chemistry 276:12041–12048.*

16. Emerson, D. K., M. L. McCormick, J. A. Schmidt, and C. M. Knudson. 2005. Taurine monochloramine activates a cell death pathway involving Bax and caspase-9. *The Journal of Biological Chemistry 280:3233–3241.*

17. McKearn, J. P., J. McCubrey, and B. Fagg. 1985. Enrichment of hematopoietic precursor cells and cloning of multipotential B-Lymphocyte precursors. *Proceedings of the National Academy of Sciences of the United States of America 82:7414–7418.*

18. Nunez, G., M. Seto, S. Seremetis, D. Ferrero, F. Grignani, S. J. Korsmeyer, and R. Dalla-Favera. 1989. Growth- and tumor-promoting effects of deregulated BCL2 in human B-lymphoblastoid cells. *Proceedings of the National Academy of Sciences of the United States of America 86:4589–4593.*

19. Scaduto, R. C., Jr., and L. W. Grotyohann. 1999. Measurement of mitochondrial membrane potential using fluorescent rhodamine derivatives. *Biophysical Journal 76:469–477.*

20. Smiley, S., M. Reers, C. Mottola-Hartshorn, M. Lin, A. Chen, T. Smith, G. Steele, Jr., and L. Chen. 1991. Intracellular heterogeneity in mitochondrial membrane potentials revealed by a J-aggregate-forming lipophilic cation JC-1. *Proceedings of the National Academy of Sciences of the United States of America 88:3671–3675.*

21. Martin, S. J., C. P. Reutelingsperger, A. J. McGahon, J. A. Rader, R. C. van Schie, D. M. LaFace, and D. R. Green. 1545. Early redistribution of plasma membrane phosphatidylserine is a general feature of apoptosis regardless of the initiating stimulus: inhibition by overexpression of Bcl-2 and Abl. *The Journal of Experimental Medicine 182:1545–1556.*

22. Panaretakis, T., K. Pokrovskaja, M. C. Shoshan, and D. Grander. 2002. Activation of Bak, Bax, and BH3-only proteins in the apoptotic response to doxorubicin. *The Journal of Biological Chemistry 277:44317–44326.*

23. Rathmell, J. C., C. J. Fox, D. R. Plas, P. S. Hammerman, R. M. Cinalli, and C. B. Thompson. 2003. Akt-directed glucose metabolism can prevent Bax conformation change and promote growth factor-independent survival. *Molecular and Cellular Biology 23:7315–7328.*

24. Mandic, A., K. Viktorsson, M. Molin, G. Akusjarvi, H. Eguchi, S. I. Hayashi, M. Toi, J. Hansson, S. Linder, and M. C. Shoshan. 2001. Cisplatin induces the proapoptotic conformation of Bak in a delta MEKK1-dependent manner. *Molecular and Cellular Biology 21:3684–3691.*

10

In Vitro and *In Vivo* Apoptosis Detection Using Membrane Permeant Fluorescent-Labeled Inhibitors of Caspases

Brian W. Lee, Michael R. Olin, Gary L. Johnson, and Robert J. Griffin

Summary

Apoptosis detection methodology is an ever evolving science. The caspase family of cysteine proteases plays a central role in this environmentally conserved mechanism of regulated cell death. New methods that allow for the improved detection and monitoring of the apoptosis-associated proteases are key for further advancement of our understanding of apoptosis-mediated disease states such as cancer and Alzheimer's disease. From the use of membrane permeant fluorescent-labeled inhibitors of caspases (FLICA) probe technology, we have demonstrated their successful use as tools in the detection of apoptosis activity within the *in vitro* and *in vivo* research setting. In this chapter, we provide detailed methods for performing *in vitro* apoptosis detection assays in whole living cells, using flow cytometry, and 96-well fluorescence plate reader analysis methods. Furthermore, novel flow cytometry-based cytotoxicity assay methods, which incorporate the FLICA probe for early apoptosis detection, are described. Inclusion of this sensitive apoptosis detection probe component into the flow-based cytotoxicity assay format results in an extremely sensitive cytotoxicity detection mechanism. Lastly, in this chapter, we describe the use of the FLICA probe for the *in vivo* detection of tumor cell apoptosis in mice and rats. These early stage *in vivo*-type assays show great potential for whole animal apoptosis detection research.

Key Words: FLICA; *in vivo*; *in vitro*; apoptosis; caspase; tumor imaging; detection; cytotoxicity assay.

From: *Methods in Molecular Biology, vol. 414: Apoptosis and Cancer*
Edited by: G. Mor and A. B. Alvero © Humana Press Inc., Totowa, NJ

1. Introduction

Apoptosis is a genetically coded, evolutionarily conserved form of cell suicide essential for the development and homeostatic maintenance of the multicellular organism. Apoptosis is characterized as a caspase-mediated process exhibiting the classic morphological features of an apoptotic cell *(1)*. These features usually consist of cytoplasmic and nuclear condensation, DNA fragmentation, membrane blebbing, and the eventual compartmentalization of the cellular cytoplasmic contents into membrane-enclosed apoptotic bodies *(2)*. The caspase family of cysteinyl aspartate-specific proteases lies at the center of this highly regulated cell death process. Consequently, they have been extensively studied and the subject of numerous publications and review articles *(3–8)*.

Caspases are initially synthesized and stored as inactive pro-caspase precursor molecules known as zymogens. Upon receiving an apoptotic signal through a number of stimuli, including granzyme B incorporation from natural killer (NK) or cytolytic T-cell attack, chemotherapeutic drugs, or environmental condition changes, the inactive pro-form of the caspases is rapidly converted into the active enzyme structure *(8)*. This process is accomplished through cleavage of the pro-domain at specific aspartic acid residues to yield large (20 kDa) and small (10 kDa) subunits that assemble into the heterotetrameric, catalytically active form of the caspase enzyme *(4,9)*. Active caspase enzymes recognize and target short tetra-peptide amino acid sequences containing an aspartate in the P1 position of the four amino acid sequence *(10,11)*. These preferred tetra-peptide sequences are located within numerous other intracellular proteins such as, poly ADP-ribose polymerase (PARP), the pro-apoptotic Bcl-2 family member Bid, and other pro-caspase molecules. This target sequence subjects these proteins to the caspase-mediated proteolytic process *(12)*.

Peptide-based inhibitors have been derived from the preferred tetra-peptide caspase target sequences that were derived from Positional Scanning Combinatorial Library studies *(11)*. Peptide-based inhibitors have been used extensively to study the specific mechanisms involving apoptotic processes and to assess the effects of apoptosis inhibition in numerous patho-physiological disease states *(13–20)*. Peptide-based inhibitor probes were especially important in the early research in caspase discovery and isolation. For example, the inhibitor probes Ac-Tyr-Val-Ala-Asp-COCHN$_2$ (Ac-YVAD-AMK) and Ac-YVAD-CHO were used to identify the interleukin-1 (IL-1) converting enzyme that is now commonly referred to as caspase 1 *(10)*. Peptide-based inhibitor probes containing a fluoromethyl ketone (FMK) functional group provide an irreversible linkage and permanent inactivation of the cysteine protease enzyme *(21)*. FMK-type O-Me (methylated) inhibitor probes exhibit high-affinity transmembrane kinetics allowing for rapid association with the caspase reactive sites *(13)*. Once this association step has been success-fully completed, a second reaction occurs in which the reactive cysteine-SH is alkylated, forming a thioether adduct *(22,23)* (*see* **Fig. 1**). These FMK probes have an increased specificity compared with the chloromethyl ketone (CMK) predecessors that bind and inactivate either cysteine or serine proteases, regardless of the tetrapeptide sequence associated with them *(22)*.

Successful use of the FMK-labeled tri- and tetra-peptide caspase inhibitors led to the development of cell permeant fluorescent-tagged caspase detection probes by Immunochemistry Technologies (Bloomington, MN). The technology was later verified by the Brander Cancer Institute (Hawthorne, NY) *(24–31)*. Detection of intracellular caspase activity is an essential element in apoptosis research. A number of approaches are currently used for analyzing the various mediators of this apoptosis-driven protease cascade, which include but are not limited to, fluorescence-labeled inhibitors of caspases (FLICA) *(26)*, fluoro-genic/chromogenic caspase substrate assays *(32–37)*, immunocytochemical detection assays that can identify the active caspase enzymes or their specific cleavage products *(38)*, DNA fragmentation detection assays through the TUNEL assay system *(39)*, and mitochondrial potentiometric dye-based assays *(40,41)*.

This chapter describes *in vitro* and *in vivo* apoptosis detection procedures focusing exclusively on the use of the membrane permeant FLICA probe methodology. Methods describing its use in in vitro apoptosis detection assays, flow cytometry-based cytotoxicity assays, and *in vivo* apoptosis detection applications are provided. *In vitro* caspase detection assays employ the cell permeant fluorescent-tagged inhibitor properties of the FLICA probes to detect up-regulated caspase activity in apoptotic adherent- and suspension-type cell cultures. The total cytotoxicity assay incorporates a FLICA probe with fluorescent general membrane and DNA-type vital stains to detect both early and late apoptotic and necrotic cells in cell-mediated cytotoxicity (CMC) activity assays. The increased sensitivity of the assay will allow it to replace the radioactive chromium (^{51}Cr) and enzyme release assays currently being used to quantitate the level of cellular mortality in various CMC and (other) experi-mental conditions. *In vivo* caspase detection assays involve the injection of the FLICA reagent directly into the circulatory system or other internal regions of the animal. Apoptotic regions showing elevated concentrations of the FLICA probe can subsequently be monitored using various detection technologies.

1.1. Intracellular Caspase Detection in Adherent and Suspension Cells

Apoptosis-associated up-regulation of the caspase cascade can be detected in whole living cells using the FLICA probe assay technique. Following exposure of the experimental cell population to various apoptosis-inducing conditions, these cells along with an identical (non-induced) control cell population are incubated with the FLICA probe. The cell permeant FLICA probe enters both experimental and control cell populations. Up-regulated caspase enzymes in apoptotic cells are covalently labeled with the FLICA probes (*see* **Fig. 1**). Following several wash steps, which remove the unbound FLICA probe from non-apoptotic cells, cells may be visualized and quantitated using fluorescence microscopy, fluorescence plate reader, and flow cytometry methods. Non-apoptotic cells should exhibit very low levels of bound fluorescence emission.

1.2. Flow Cytometry-Based Total Cytotoxicity Assay

Detection and quantitation of cellular death is an important and essential process for understanding the pathogenic mechanisms associated with many

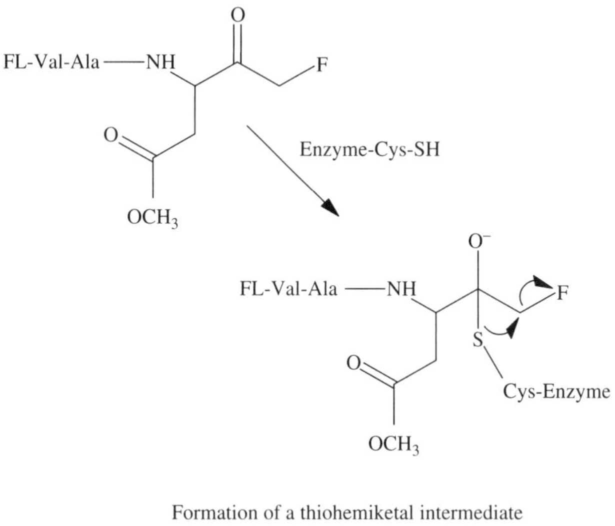

Formation of a thiohemiketal intermediate

Fig. 1. Covalent binding mechanism of a FAM-VAD-FMK inhibitor probe to cysteine protease enzymes that include the caspase family of apoptosis-associated proteases. Once the inhibitor probe sequence has been recognized by the caspase reactive binding site (slow reaction), then the final covalent binding step (fast reaction) proceeds almost immediately.

intracellular bacterial and viral pathogens. A determination of the degree of cell death is central to the study of graft rejection in organ transplantation research. Although there are a number of different types of assays for the detection of cellular death, CMC is most often measured using the ^{51}Cr release assay, first described by Brunner et al. *(42)*. Assays of this type have been used to study a wide range of CMC-driven processes. A few of these CMC-driven events include antibody-dependent CMC of HIV-infected cells *(43)*, cytotoxic T lymphocyte targeting of virus-infected cells *(44)*, cell targeting of melanoma cancer cells *(45)*, graft versus host disease *(46)*, and T-cell-mediated insulin-dependent diabetes *(47)*.

^{51}Cr release assays are based on the passive internalization and binding of ^{51}Cr from sodium chromate within the target cells. Lysis of the target cells by CMC results in the release of the radioactive probe into the cell-culture supernatant *(42)*. Although ^{51}Cr release assays can provide useful quantitative information regarding the level of CMC present in the cell population(s), major concerns have been raised about the high cost of running the assays and radiation exposure issues for laboratory workers *(48)*. Other issues include the disposal of radioactive reagent and waste products, the detrimental effect of radioactivity emission on cell function *(49)*, a labor-intensive protocol, wide variations in radioactive labeling, and a relatively high level of spontaneous ^{51}Cr release *(50)*.

Over the past 10 years, the flow cytometer has rapidly become a powerful tool for analyzing cellular function. The flow cytometer is a laser-based instrument capable of analyzing cells and their products through the excitation of fluorescent tags conjugated to antibodies or probes. Flow cytometry-based cytotoxicity assays usually include a fluorescent membrane stain to label and differentiate target cells from effector cells plus a fluorescent vital stain to detect the dead and dying target cell population *(50–55)*. Flow cytometric assays of this design show an excellent correlation with the ^{51}Cr release assays that were run in parallel for comparison. Typical r^2 linear regression analysis values ran from 0.960 to 0.982 when percentage of cytotoxicity values from the two assay formats were compared *(50,51,53)*. Flow cytometric assays of this design have a number of advantages over the traditional ^{51}Cr release assays. These include ability to monitor cytotoxicity effects on a single-cell level, detection of early apoptosis (increased sensitivity), as well as necrotic cells, no radiation hazard, elimination of detrimental radiation effects on the effector population, shorter assay turnaround times, lower cost, ability to differentiate subpopulations of cells (e.g., live targets, killed targets, live effectors, and dead effector cells), and allow the assessment of the effector cell population viability *(50–52,56)*.

Flow cytometry-based cytotoxicity assays of the type in the previous paragraph allowed for an easy and accurate examination of cellular death (*50–55*). However, all single-sampling assay formats rely on a single time point for their estimation of the cytotoxicity status of the target sample population. The dependence on a cell membrane integrity-based fluorescent vital dye as the only means by which to determine the percentage of cells entering cellular death through apoptosis is inadequate. Early stage apoptotic cells retain intact membrane integrity and are vital stain negative. Although these cells will eventually become cell death positive by the membrane integrity (vital staining) detection method, at the time of measurement, they will erroneously be labeled as negative healthy cells. This leads to a substantial underestimation of the actual degree of cell death activity from the CMC-associated interaction event. Incorporation of a cell permeant caspase detection probe, such as the FLICA reagent, into the flow cytometry-based cytotoxicity detection assay, allows for the additional quantitation of the previously undetectable early apoptotic cell population. Because the FLICA probe will readily penetrate the cell membrane without permeabilization, cells entering into early stage apoptosis are accurately detected (*see* **Fig. 2**).

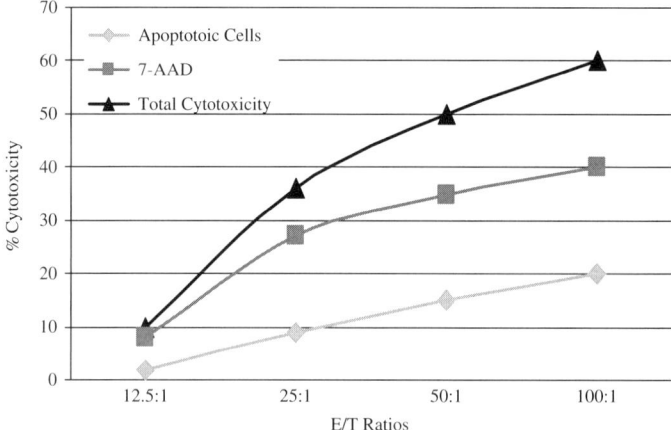

Fig. 2. Percentage of cytotoxicity detected by apoptosis detection probe (SR-VAD-FMK) only, 7-AAD only, and combined 7-AAD and apoptosis probe analysis methods. Combining an apoptosis detection probe with the 7-AAD vital stain results in a cytotoxicity assay with greater sensitivity.

1.3. In Vivo Apoptosis Detection Assay

Fluorescent-labeled FMK peptides, such as carboxyfluorescein-Val-Ala-Asp(O-Me)-FMK (FAM-VAD-FMK) and sulforhodamine-B-Val-Ala-Asp(O-Me)-FMK (SR-VAD-FMK), can be used to detect apoptosis *in vivo*. The cell permeant FLICA probes can be injected intravenously (IV) into live mice, allowed to circulate in the animal, and label areas of apoptosis. These areas can be visualized through observation of the localization of the fluorescence signal in areas having elevated levels of apoptotic activity.

2. Materials

2.1. In Vitro Apoptosis Detection Assays Using Flow Cytometry and 96-Well Fluorescence Plate Reader Formats

2.1.1. Equipment

1. Hemacytometer (VWR Scientific, West Chester, PA) or other methods.
2. Flow cytometer equipped with a single 488-nm argon laser and 530-, 585-, and >650-nm filters for analysis in FL1, FL2, and FL3 channels, respectively.
3. 96-well fluorescence plate reader with 488–492 nm excitation and 520–530 nm emission filter capability.

2.1.2. Non-Fluorescent Reagents or Assay Components

1. Phosphate-buffered saline (PBS), pH 7.4: prepare in 500 ml quantities and store at room temperature: 140 mM NaCl, 2.7 mM KCl, 8 mM Na_2HPO_4, and 1.5 mM KH_2HPO_4 (dry reagents from J.T. Baker, Phillipsburg, NJ).
2. Bovine serum albumin (BSA) (Sera Care Life Sciences, Oceanside, CA).
3. Apoptosis wash buffer (AWB), pH 7.4: dissolve BSA powder into PBS buffer to obtain a 5 mg/ml BSA concentration. Sodium azide (J.T. Baker), at a concentration of 0.05%, can be added for long-term storage at 2–8°C.
4. Cell-culture media, RPMI-1640 (Atlanta Biological, Lawrenceville, GA), containing 10% FBS, 1 mM sodium pyruvate, non-essential amino acids, 2 mM L-glutamine, 10 mM HEPES, 100 U/ml penicillin, and 100 µg/ml streptomycin (all media additives obtained from Atlanta Biological).
5. Staurosporine (LKT Labs, St. Paul, MN).
6. Camptothecin (LKT Labs).
7. Trypsin-versene (Cambrex, Walkersville, MD).
8. Black 96-well microtiter plates (Corning-Costar, Corning, NY).
9. Molecular grade DMSO (Sigma Aldrich, St. Louis, MO).
10. Jurkat cells (ATCC, Manassas, VA).

2.1.3. Apoptosis Detection—FAM-VAD-FMK and FAM-DEVD-FMK (FLICA) Probes

1. Reconstitute vials of lyophilized FAM-VAD-FMK and FAM-Asp-Glu-Val-Asp-FMK (FAM-DEVD-FMK) (Immunochemistry Technologies, Bloomington, MN) with 50 µl DMSO (*see* **Note 1**). Mix/vortex vigorously. This yields a FLICA probe concentration of (1 mg/ml) and forms a 150× stock concentrate.
2. Just prior to use, add 200 µl PBS to the vial and mix/vortex vigorously. This produces a 30× FLICA probe concentration, which should be added to cell cultures at a 1:30 v/v ratio as quickly as possible (*see* **Note 2**).

2.1.4. 7-AAD Vital Stain

1. Reconstitute lyophilized 7-aminoactinomycin D (7-AAD) (AnaSpec, San Jose, CA) with DMSO to a concentration of 1 mg/ml yielding a 200× stock concentrate.
2. Using PBS, make a 1:10 dilution of 7-AAD yielding a 20× (0.1 mg/ml) solution.
3. Add to cell cultures at a 1:20 v/v ratio just prior to reading on the flow cytometer. 7-AAD-stained cells may be kept for longer periods (30 min) if stored on ice and protected from light.

2.2. Total Cytotoxicity Assay

2.2.1. Equipment

1. Flow cytometer equipped with a single 488-nm argon laser and 530-, 585-, and >650-nm filters for analysis in FL1, FL2, and FL3 channels, respectively.

2.2.2. Non-Fluorescent Reagents

1. PBS, pH 7.4: prepare in 500 ml quantities and store at room temperature: 140 mM NaCl, 2.7 mM KCl, 8 mM Na_2HPO_4, and 1.5 mM KH_2HPO_4 (dry reagents from Mallinckrodt Baker, Phillipsburg, NJ).
2. Cell-culture media, RPMI-1640 (Atlanta Biological) containing 10% FBS (Atlanta Biological).
3. Molecular grade DMSO (Sigma Aldrich).

2.2.3. Target Cell Stain—CFSE

1. Dissolve 5(6)-FAM diacetate, succinimidyl ester (CFSE) (AnaSpec) with DMSO to a 2500× working concentration (250 µg/ml).
2. Further dilute CFSE 1:250 in sterile PBS to make a 10× concentration.

2.2.4. Apoptosis Detection—SR-VAD-FMK

1. Reconstitute lyophilized SR-VAD-FMK (Immunochemistry Technologies) with DMSO to a concentration of (2.62 mg/ml) yielding a 252× stock concentrate.
2. Using RPMI-1640, make a 1:12 dilution SR-VAD-FMK yielding a 21× solution.

2.2.5. Vital Stain — 7-AAD

1. Reconstitute lyophilized 7-AAD (AnaSpec) with DMSO to a concentration of 1 mg/ml yielding a 210× stock concentrate.
2. Using PBS, make a 1: 10 dilution of 7-AAD yielding a 21× solution.

2.3. In Vivo Apoptosis Detection Assay

2.3.1. Equipment

1. Flow cytometer equipped with a single 488-nm argon laser and 530-, 585-, and >650-nm emission filters for analysis in FL1, FL2, and FL3 channels, respectively.

2.3.2. Reagents

1. Green FLIVO™ reagent, FAM-VAD-FMK, 50 μg per vial (Immunochemistry Technologies). Dissolve in 50 μl DMSO. Dilute by adding 200 μl sterile PBS, pH 7.4.
2. Red FLIVO™ reagent, SR-VAD-FMK, 131 μg per vial (Immunochemistry Technologies). Dissolve in 50 μl DMSO. Dilute by adding 200 μl sterile PBS, pH 7.4.
3. Arsenic trioxide (ATO or Trisenox, Cell Therapeutics, Seattle, WA), use 8 mg/kg for intraperitoneal (IP) injection.

3. Methods

3.1. Methods for Performing the 96-Well and Flow Cytometer Assay

3.1.1. Cell Culture

1. Cultivate Jurkat cells in cell-culture media taking care not to allow cultures to become too dense and overgrown (>10^6 cells/ml). Cell concentrations may be determined by counting in a hemacytometer (*see* **Note 3**). If adherent cells are to be used, they should be seeded at a low concentration (<1×10^5 cells/ml) onto the surface of the chamber that the reaction will be performed in and grown to about 80–90% confluence (*see* **Note 4.**)
2. Split the cell-culture flask contents 1–2 days prior to performing the apoptosis induction procedure. This serves to replace depleted media as well as adjust the cell concentration to an optimal concentration density ($3–6 \times 10^5$ cells/ml) (*see* **Note 5**).

3.1.2. Induction of Apoptosis in Experimental Cell Population

1. Concentrate the cell-culture suspensions by centrifugation and recon using fresh cell-culture media. Use centrifugation speeds yielding <$500 \times g$. When performing the 96-well fluorescence plate reader assay, cells should be adjusted to a 3×10^6 to

1×10^7 cells/ml concentration. Flow cytometry analysis can be performed at 5- to 10-fold lower cell concentrations because of the nature of the detection system.

2. Set up the proper control conditions for performing flow cytometry and 96-well fluorescence plate reader analysis. When performing Bicolor-Staining Flow Cytometry analysis, three types of staining conditions are recommended for the establishment of proper electronic compensation and quadrant statistics. They consist of (i) cells stained with FLICA only (induced and non-induced), (ii) cells stained with 7-AAD only (induced and non-induced), and (iii) cells stained with FLICA and 7-AAD (induced and non-induced). Fluorescence plate reader analysis does not require more than the inclusion of known induced (positive) and non-induced (negative) cell populations to validate the utility of the FLICA probe analysis.

3. Expose the experimental cell population to our particular apoptosis-inducing agent. For illustrative purposes, we will use a 1 μM staurosporine (0.466 μg/ml) apoptosis-inducing condition (*see* **Note 6**).

4. Incubate Jurkat cell apoptosis induction model at 37°C for 3–6 h.

5. Simultaneously, incubate an identical population of Jurkat cells with a concentration of DMSO (carrier) that is identical to that which is present in the induced cell-culture flasks. This condition will serve as a negative control for these apoptosis induction assays (*see* **Note 7**).

3.1.3. Incubation of Cell Cultures with the FLICA Reagent

1. Add reconstituted FLICA reagent at either the 150× or the 30× concentration, depending on volume of cell suspension (pipetting precision purposes) that is to be analyzed. In this method procedure, we used a FAM-VAD-FMK probe in the 96-well fluorescence plate reader method and a FAM-DEVD-FMK probe for the flow cytometry detection assay. FLICA reagent should be added to both the induced and the non-induced cell populations.

2. Gently mix the FLICA reagent in the cell media to assure a homogeneous solution. Incubate for 1 h in a 37°C CO2 incubator, taking care to occasionally mix suspension cultures every 20–30 min.

3.1.4. Cell-Culture Wash Procedure

1. Wash the FLICA reagent treated cell cultures using the AWB. For suspension cell cultures, transfer cells to 15-ml sterile polypropylene tubes and add a 1:5 v/v aliquot of the AWB to each tube. Mix well. When washing adherent cells, carefully decant the supernatants from the plate-well surface, taking care not to pull up any non-adherent, apoptotic cells from the cell surface.

2. Centrifuge ($<500 \times g$) to pellet suspension cells and remove and discard supernatants.

3. Add another 1–2 ml AWB to either the suspension cell pellets or the adherent cell chambers. When washing adherent cells (suspension cells too), allow at least 3–5 min of AWB exposure time for each wash step (*see* **Note 8**).

4. Perform the wash steps two more times to remove any non-specifically bound FLICA reagent from non-induced cells.

5. Resuspend suspension cells in 0.5–1.0 ml AWB. Note, when using 96-well fluorescence plate reader analysis methods, it is important to assure that the number of induced and non-induced cells/well is approximately identical. To count cells, pull out a small aliquot of each induction status flask and count in a hemacytometer. Adjust cell concentrations by adjusting resuspension volumes of either the induced or the non-induced sample tubes. This cell concentration equalization step is not necessary when performing flow cytometry analysis.

6. Transfer 100 μl aliquots to black microtiter plates for analysis. Adherent cells may be disassociated from the cell chamber surface by incubation with trypsin-versene solution. Once cells become detached from the chamber surface, they can be removed and suspended in cell-culture media containing FBS. Following a brief centrifugation (<500 × *g*), they can be resuspended in a reduced volume of AWB and examined using either flow cytometry or 96-well plate reader analysis methods.

3.1.5. Apoptosis Detection and Analysis: Flow Cytometry Method

1. Stain the (7-AAD only) flow cytometer control pair (induced and non-induced) described in **Subheading 3.1.2**, step 2, for 7-AAD compensation by diluting the 20× 7-AAD stain at a 1:20 v/v ratio to the volume of cell suspension present in the control tubes.

2. Stain a second control pair (induced and non-induced, FLICA, and 7-AAD stained controls) with 7-AAD (*see* **Subheading 3.1.2.**, step 2). In this control pair, tubes containing FLICA-stained Jurkat cells are stained with 7-AAD using a 1.20 v/v ratio of 7-AAD (20×) concentrate to Jurkat cell suspension volume.

3. Set aside a third control tube pair that was stained with the FLICA reagent only (*see* **Subheading 3.1.2.**, step 2).

4. Incubate 7-AAD samples for 5 min on ice in dark.

5. Read samples on flow cytometer within 20 min.

6. Set up flow cytometer compensation settings to allow the clear differentiation of the green fluorescent both induced (positive) and non-induced (negative) cell populations and the differentiation of the healthy non-apoptotic 7-AAD-negative cell population from the membrane compromised (dead and dying) 7-AAD-positive cell population. For bicolor analysis, measure fluorescein fluorescence in the FL1 channel and the red fluorescence (7-AAD) in the FL3 channel.

7. Generate a log FL1 (X-axis) versus log FL3 (Y-axis) dot plot.

8. Put in quadrant cursors. The four quadrant areas contain the following cell populations: (i) quadrant 1, 7-AAD-positive/fluorescein-negative cells; (ii) quadrant 2,

fluorescein/7-AAD-positive cells; (iii) quadrant 3, fluorescein/7-AAD-negative cells; and (iv) fluorescein-positive/7-AAD negative cells (*see* **Fig. 3** and **Note 9**).

3.1.6. Apoptosis Detection and Analysis: Fluorescence Plate Reader Method

1. Turn on 96-well plate reader to allow unit to warm up and equilibrate.
2. Select the proper read-out settings for the particular make and model of fluorescence plate reader in our laboratory. In this example, we used a Molecular Devices Spectra-Max Gemini fluorescence plate reader equipped with adjustable wavelength excitation and emission features for performing endpoint, kinetic, spectrum, and well-scanning readings. On our instrument, we set up the reader to excite the microtiter plate-well samples at 488 nm, read the fluorescence emissions at 530 nm, and use a 515 nm emissions cut-off filter setting to reduce plate and excitation noise.
3. Transfer an equal number of washed, FLICA probe-treated (induced and non-induced) Jurkat cell suspensions into black microtiter plate wells. Read at least three 100 μl/well replicates per induced versus non-induced pairing (*see* **Note 10**).
4. Calculate the average random fluorescence units (RFUs) output from each set of replicate wells and record in a laboratory notebook.

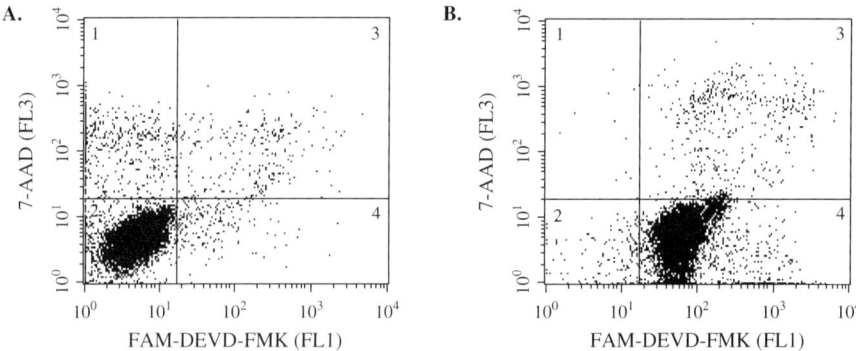

Fig. 3. Scatter plot of staurosporine-induced Jurkat cells dual stained with 7-AAD (vital stain) and FAM-DEVD-FMK apoptosis detection probe. Jurkat cells were incubated at 37°C for 3.5 h in cell-culture media containing DMSO carrier (**A**) or 1 μM staurosporine (**B**). Following an hour incubation with FAM-DEVD-FMK and subsequent wash steps, cells were stained with 7-AAD and analyzed on the flow cytometer using the FL1 (fluorescein) and FL3 (7-AAD) channels. Quadrant 1, = caspase-negative, dead/necrotic cells. Quadrant 2, living, non-apoptotic cells. Quadrant 3, dead and dying, caspase-positive cells. Quadrant 4, early apoptotic cells with intact membrane integrity.

5. Express induced versus non-induced RFUs output data in terms of an induced/non-induced (I/NI) output ratio. For any given apoptosis induction experiment, the ratio of induced population fluorescence over the non-induced (control) population fluorescence is directly proportional to the degree of apoptosis induction in that particular cell line using that specific apoptosis-inducing mechanism.

6. Create a bar graph to illustrate the differential RFUs output readings obtained from a FLICA reagent probe analysis of an induced and non-induced cell population. In this example, we show the results of a FAM-VAD-FMK caspase detection probe analysis of a set of six 5-h staurosporine-treated Jurkat cell cultures (*see* **Fig. 4** and **Note 11**).

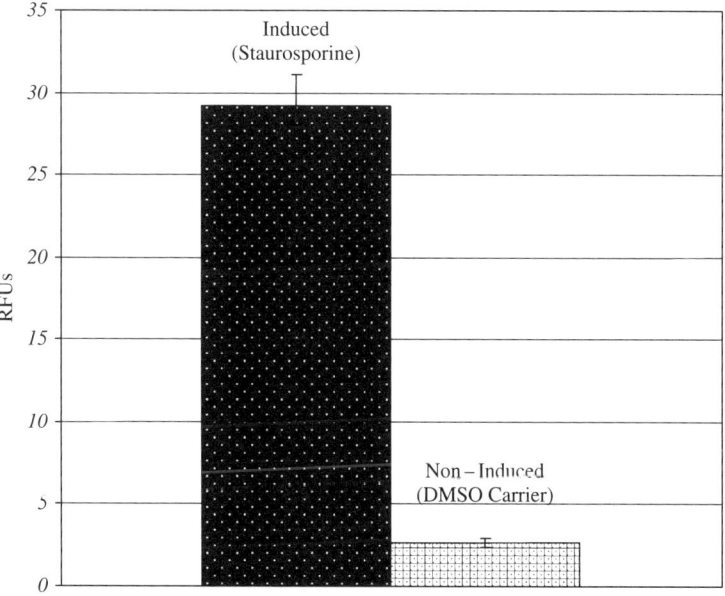

Fig. 4. 96-well fluorescence plate reader analysis of apoptotic Jurkat cells using FAM-VAD-FMK (FLICA) reagent probe. Jurkat cells were exposed to 1 μM staurosporine or DMSO carrier control for 5 h at 37°C, incubated for 1 h at 37°C with 10 μM FLICA reagent, washed 3× in apoptosis wash buffer (AWB), and analyzed in a Molecular Devices Spectra-Max 96-well fluorescence plate reader using 488 nm excitation/530 nm emission settings. The black (induced) and white (non-induced) columns represent the mean RFUs output values of a total of six separate staurosporine induction assays that were probed using the FLICA reagent. The RFUs output for any single induction assay is the sum of three 100 μl/well replicates.

3.2. Methods for Performing the Total Cytotoxicity Assay

3.2.1. Assay Controls

1. CFSE-stained target cells + effector cells.
2. CFSE + 7-AAD-stained target cells.
3. CFSE-stained target cells + 7-AAD + effector cells.
4. CFSE + SR-VAD-FMK-stained target cells.
5. CFSE-stained target cells + SR-VAD-FMK + 7-AAD.
6. CFSE-stained target cells + SR-VAD-FMK + 7-AAD + effector cells.

3.2.2. Assay Procedure

1. Wash target cells (monocytes, K562 cells, or other antigen-presenting cell) 2× in 1× PBS.
2. Resuspend approximately 1–2×10^7 target cells in 1.8 ml PBS.
3. Add 200 µl 10× CFSE (*see* **Subheading 2.2.3.**, step 2), vortex well, and incubate at room temperature for 15 min.
4. Add 2 ml cell-culture media, centrifuge at $<500 \times g$ for 5 min, and discard supernatant.
5. Resuspend cells in 4 ml cell-culture media and incubate in 37°C incubator for 30 min.
6. Centrifuge at $<500 \times g$ for 5 min, discard supernatant, and resuspend cells in 200–500 µl cell-culture media to concentrate the cells.
7. Prepare 100 µl aliquots containing 3×10^4 target cells (in cell-culture media) in sterile FACS tubes.
8. Add effector cells (PBMCs, CD8, CD4, γδ, etc.) at a concentration to achieve a desired effector/target (E/T) cell ratio (*see* **Note 12**) in a total volume of 200 µl (*see* **Fig. 5**). Add 100 µl media to each of the control tubes.
9. Incubate for 4–6 h. Incubation time will vary according to experimental design and cell type.
10. Add 10 µl 20× SR-VAD-FMK (*see* **Subheading 2.2.4.**, step 2) to all samples approximately 45 min before the end of the 4- to 6-h incubation period (do not wash).
11. Following incubation, add 200 µl cell-culture media to all samples.
12. Place all samples on ice.
13. Add 20 µl 21× 7-AAD (*see* **Subheading 2.2.5.**, step 2) to appropriate control tubes.
14. Set up flow cytometer (*see* **Subheading 3.2.3.**).
15. Following the set up of the flow cytometer (*see* **Subheading 3.2.3.**), add 7-AAD to the remaining test samples and gently vortex.
16. Incubate samples for 5 min on ice in dark (*see* **Note 13**).
17. Immediately read samples on flow cytometer or place back on ice.

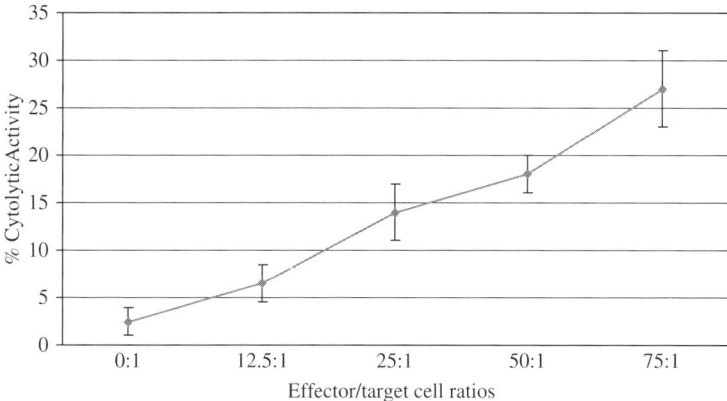

Fig. 5. Plot of effector/target (E/T) cell ratios versus percentage of target cell killing. Target cells were incubated with different E/T ratios to determine the optimal E/T ratio for maximal cytotoxicity induction.

3.2.3. Flow Cytometry Set Up

1. Resuspend approximately 1×10^6 target and effector cells (*see* **Subheading 3.2.1.**, control 1) at a 1:2 E/T cell ratio to set up an side scatter (SS) versus CFSE plot. Draw a gate around the CFSE-positive target cell population (*see* **Fig. 6**). If greater separation is required, incubate target cells with a higher concentration of CFSE.
2. Create a CFSE versus 7-AAD plot gated from target population in **Fig. 6** and run CFSE- and 7-AAD-stained target cell control (*see* **Subheading 3.2.1.**, control 2) (*see* **Fig. 7**).
3. Using CFSE versus 7-AAD plot, run CFSE- and 7-AAD-stained target cells + effector cells control tube (*see* **Subheading 3.2.1.**, control 3) to ensure adequate separation between effector and target cells. Draw a gate around target cell population (*see* **Fig. 8**). Set flow cytometer to collect 5000 events from the gate created in **Fig. 8**.
4. Create a 7-AAD versus SR-VAD-FMK plot gating on the target population that was gated in **Fig. 8** and rerun:
   a. CFSE-stained target cells + SR-VAD-FMK-stained control (*see* **Subheading 3.2.1.**, control 4) ensuring proper compensation,
   b. CFSE-stained target cells + 7-AAD-stained control (*see* **Subheading 3.2.1.**, control 2) ensuring proper compensation, and
   c. CFSE-stained target cells + SR-VAD-FMK + 7-AAD-stained control (*see* **Subheading 3.2.1.**, control 5) (*see* **Fig. 9**).

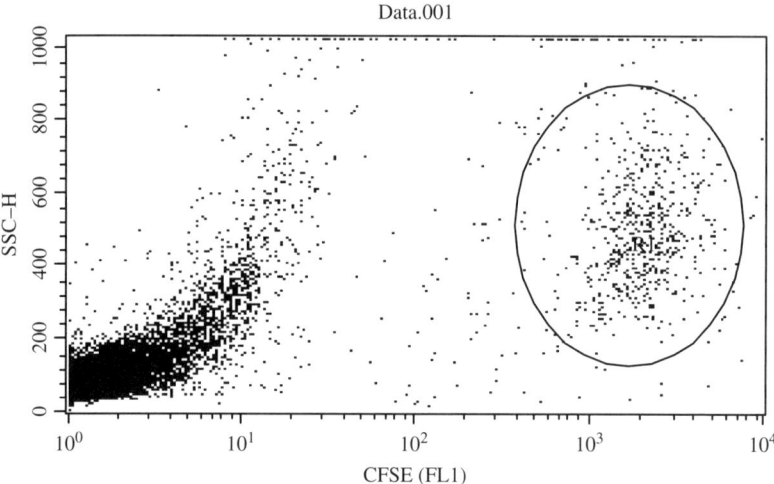

Fig. 6. SS versus CFSE plot of an unstained effector and CFSE-stained target cell population. Staining of target cells with the membrane stain, CFSE, allows for an easy separation of the effector and target cell populations. Gating on these CFSE-stained target cells allows for subsequent analysis using SR-VAD-FMK caspase detection probe and the 7-AAD vital stain.

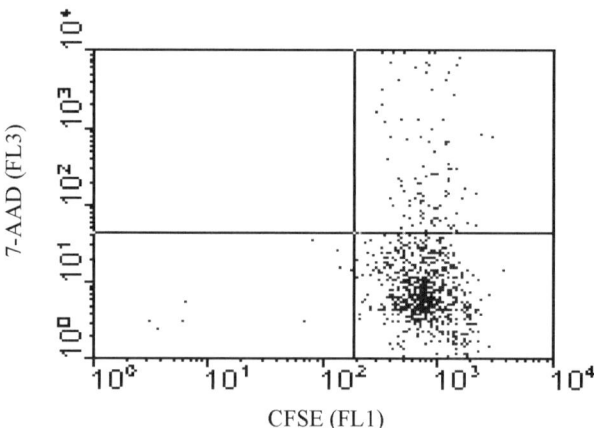

Fig. 7. CFSE versus 7-AAD plot of target cells stained with CFSE membrane stain and 7-AAD vital stain. The PMT voltage was adjusted to place the stained target cells in the third or fourth log decade.

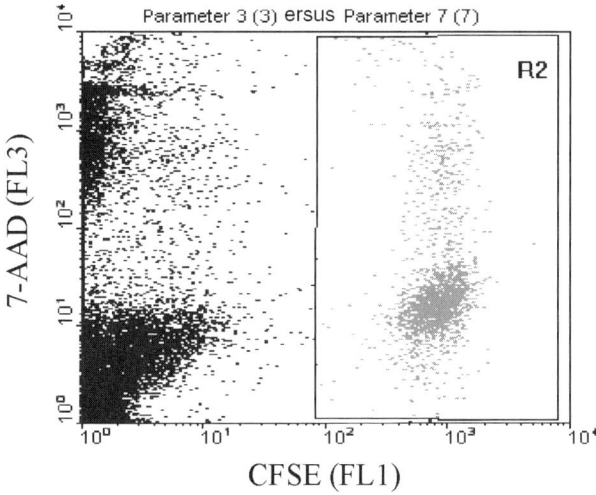

Fig. 8. CFSE versus 7-AAD plot of target + effector cell populations stained with CFSE and 7-AAD. This plot is run to ensure proper separation between target and effector cell populations. CFSE-stained target cells are gated within region R2 where all subsequent analysis will take place.

5. Run CFSE-stained target cells + SR-VAD-FMK stain + 7-AAD stain + effector cells control (*see* **Subheading 3.2.1.**, control 6) to ensure all parameters are set up correctly.

3.2.4. Total Cytotoxicity Calculation

1. Using the 7-AAD versus SR-VAD-FMK plot gated on the target population in **Fig. 8**, the total percentage of cytotoxicity from this effector cell interaction can be defined as the sum of the events in quadrants 1, 3, and 4 divided by the total number of events in all four quadrants × 100 (*see* **Fig. 9** and **Note 14**).

3.3. Methods for Performing the In Vivo Apoptosis Detection Assay

3.3.1. Assay Controls

1. Include a set of mice that receive only sterile saline injections instead of ATO. These mice will still receive the same quantity of the FLIVO™ reagent consisting of either the FAM-VAD-FMK or the SR-VAD-FMK fluorescent-labeled caspase inhibitor probe. Differences in fluorescence staining because of the presence or absence of the apoptosis-inducing chemotherapeutic agent can then be documented.

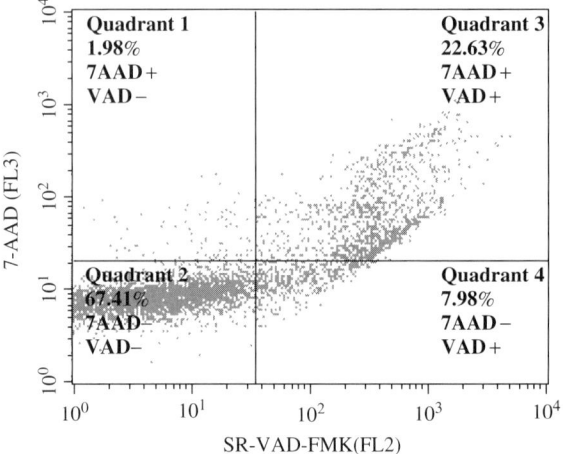

Fig. 9. SR-VAD-FMK versus 7-AAD plot showing the four different types of target cells present, following an effector CMC reaction. This analysis was performed on the CFSE-stained target cell population gated within the R2 region of the CFSE versus 7-AAD plot shown in **Fig. 8**. The total cytotoxicity percentage from this effector cell interaction can be defined as the sum of the events in quadrants 1, 3, and 4 divided by the total number of events in all four quadrants × 100. Quadrant 1, caspase-negative, dead/necrotic cells. Quadrant 2, living, non-apoptotic cells. Quadrant 3, dead and dying, caspase-positive cells. Quadrant 4, early apoptotic cells with an intact membrane integrity.

3.3.2. Assay Procedure

1. 2×10^5 Song, Clement, Kang (SCK) murine mammary carcinoma cells were injected into a skin fold in a window chamber in *A/J* mice and allowed to grow for 7–9 days.
2. The mice were injected IP with either saline (controls) or ATO at 8 mg/kg (test) to induce apoptosis in the tumors.
3. Six hours after injection with saline or ATO, mice were injected IV with 40 µl FAM-VAD-FMK (8.0 µg) as prepared in **Subheading 2.3.2**, step 1, or mice were injected with 40 µl SR-VAD-FMK (21.0 µg) as prepared in **Subheading 2.3.2**, step 2.
4. The cell permeant probe, either FAM-VAD-FMK or SR-VAD-FMK, was allowed to circulate in the mouse for 30 min before analysis.

3.3.3. Fluorescence Microscopy Observation

1. Tumors in the control and test mice were examined through the window chamber using a suitable microscope capable of detecting fluorescence (*see* **Note 15**).
2. Fluorescent images are captured using a camera or fluorescent imaging device (*see* **Note 16**).

3.3.4. Flow Cytometry Analysis

1. Tumor tissue was collected from the window chamber by scraping the tumor out of the chamber into a trypsin solution, stirring for 30 min with 10 µg/ml DNase and 5 µg/ml collagenase, and finally filtering the suspension using a 70 µM cell strainer (*see* **Note 17**).
2. Flow cytometry was then performed using a FACS Caliber flow cytometer (Becton Dickinson Immunocytometry System, San Jose, CA) for the analysis of apoptosis in the cell population obtained from the window chamber. The fluorescence derived from the intracellular-bound FLICA inhibitor probes was monitored through argon laser-driven 488 nm excitation/530 nm emission filter-tandem setting.
3. All data were acquired with an event acquisition set for 10,000 events. Data were analyzed using CellQuest Pro cytometer software (*see* **Note 18**). The results of the flow cytometry analysis demonstrate that ATO induces a greater level of apoptosis in SCK tumor cells, 39% in ATO treated mouse versus 18% in saline control mouse.

4. Notes

1. The FAM-labeled FLICA reagent probes are synthesized with nine different amino acid caspase targeting sequences to accommodate the differing caspase reactive site preferences. The VAD sequence is a generic sequence that will associate with caspases 1, 2, 3, 6, 8, 9, 10, and 13 to a varying degree. Other sequences consisting of VDVAD are preferred by caspase 2, DEVD by caspase 3, VEID by caspase 6, LETD by caspase 8, LEHD by caspase 9, AEVD by caspase 10, and LEED by caspase 13. None of these preferred caspase target sequences are totally specific to any one type of caspase.
2. Once the FLICA probe reagents have been diluted into PBS to give the 30× intermediate concentrate solution, they should be further diluted into the cell cultures as soon as possible to minimize the possibility of FMK hydrolysis.
3. Cell concentrations in suspension cultures can easily be determined by counting in a hemacytometer. To perform this procedure, remove an aliquot (0.1 ml) and dilute 1:10–1:20 in PBS. Mix the cells to obtain a homogeneous suspension. Add, using a Pasteur pipette, a small drop to the grooved edge of a hemacytometer, allowing the cell suspension to be drawn under the coverslip. Count the number of cells within each of the four corners (just outside of the narrow grid lines) and determine the average cell number/corner. Multiply this average cell number by 1×10^4 to get the cells per milliliter concentration in our 1:10 or 1:20 dilution. Then multiply by our dilution factor to obtain the cell concentration in the original cell-culture flask.
4. Care must be taken to avoid spontaneous apoptosis because of adherent cell monolayer overgrowth. Ideally, adherent cell cultures should not be used for apoptosis induction studies if they have been at 100% confluence for more than a 24-h period. Usually 80–90% confluence level monolayer cultures are optimal.

5. Suspension cell cultures should not be allowed to exceed 1×10^6cells/ml. These excessive concentrations will result in nutrient depletion and cell by-product build up in the media. This will begin to initiate an increasing level of spontaneous apoptosis in the cell-culture flask, which should always be avoided.

6. Apoptosis-inducing agents and the cell lines that are exposed to them will vary extensively in their potency and cell susceptibility, respectively. It is usually necessary to establish, at least a good positive control (inducer + cell recipient) system where a known concentration of inducing agent will generate a known apoptosis induction event within a known incubation period.

7. Cells will vary in their tolerance of DMSO in the cell-culture media. If our apoptosis-inducing agent requires DMSO as a solubilizing agent or carrier, then that same volume of the DMSO carrier should be added to the non-induced control cell population so that any carrier-associated effects can be subtracted from the induced population analysis.

8. Adherent cell monolayer cultures require at least a 3- to 5-min incubation with wash buffer/wash to allow any unbound FLICA probe to diffuse back out of the cell. Failure to allow adequate wash buffer equilibration time will result in higher than expected background fluorescence in the non-apoptotic cells.

9. Partitioning the dot plots into quadrants allows for the quantitation of the four different cell population types that are present in the induced/non-induced cell-culture samples. In quadrant 1, where cells are 7-AAD positive (red) and green fluorescein (FLICA) negative, we have a necrotic and/or very late stage apoptotic cell population present. In quadrant 2 where we have very low green FLICA and red 7-AAD fluorescence staining, we have our healthy, non-induced cell population present. In quadrant 3 where we have both 7-AAD- and FLICA-positive cell fluorescence, we have the majority of our apoptotic cell population present. These cells fluoresce green from the bound FLICA reagent and red from the bound 7-AAD (vital stain dye), which was able to enter the cells because of a breakdown in the cell membrane integrity. In quadrant 4 that contains cells exhibiting low 7-AAD fluorescence and elevated FLICA green fluorescence, we have the very early apoptotic cell population present that still retain their cell membrane integrity.

10. 96-well fluorescence plate reader analysis requires a higher cell density than if reading on a flow cytometer. Ideally, cell quantities in the black microtiter plate well should range from 5×10^5 to 1×10^6 cells/well although good results can be obtained using as few as 2×10^5 cells/well. The required number is dependent on the degree and method of apoptosis induction in the particular experiment.

11. 96-well fluorescence plate reader analysis requires that multiple cell suspension aliquots be taken from the reaction tube and transferred into replicate wells on a black 96-well plate. Typically, 100 µl aliquots work best. An average can then be taken of these multiple sample readings for enhanced accuracy.

12. Titration of optimal effector/target cell ratios is often performed to define the range of maximum cytotoxicity effectiveness. This effector to target cell ratio can

then be utilized in less clearly defined experimental situations to enhance assay sensitivity.

13. 7-AAD, the red (FL3) fluorescent vital staining dye, will penetrate very quickly into membrane compromised cells and associate with the guanine (G) and cytosine (C) regions of the DNA.

14. Utilization of an apoptosis detection probe in tandem with a cell viability probe allows this flow cytometry-based total cytotoxicity determination assay to detect cell membrane intact, early to mid-apoptotic cells, which would otherwise be missed using a traditional ^{51}Cr assay method. Traditional chromium and enzyme release methods are only capable of detecting cells located in quadrants 1 and 3. This leaves (undetected) those cells in quadrant 4, which are caspase positive but which still have intact cell membrane structure. The net result of traditional cytotoxicity detection assay methods is an underestimation of the true level of cytotoxicity activity.

15. Fluorescence was detected on an Eclipse TE200 bench-top microscope (Huntley, IL) with a band pass filter (excitation at 488 nm, emission at 520 nm for the FAM-VAD-FMK and excitation at 550 nm, emission at 580 nm for the SR-VAD-FMK).

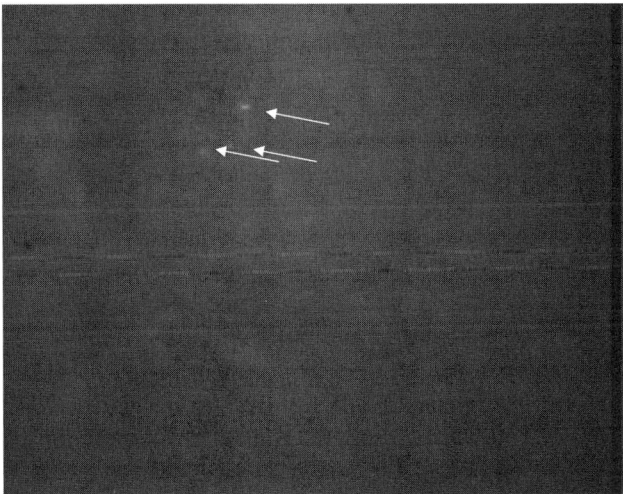

Fig. 10. *In vivo* apoptosis detection in a (saline-treated) control mouse, 7-day-old SCK tumor, following exposure to the saline control (saline IP injection). The control mouse received an IV injection of 8.0 μg cell permeant, FAM-VAD-FMK poly-caspase detection probe 6 h after saline injection. The probe was allowed to circulate for 30 min before a 20× fluorescent image was captured using a Hamamatsu C2400 camera and Broadway Imaging Software on an Eclipse TE200 bench-top microscope.

16. Fluorescent images were captured at 20× using a Hamamatsu C2400 camera (Hamamatsu Photonics, Hamamatsu City, Japan) and Broadway Imaging Software (Data Translation, Malboro, MA) on an Eclipse TE200 bench-top microscope (Nikon). The first photograph shows the negative control mouse (saline injection) (*see* **Subheading 3.3.1.**, control 1) that received an IV injection of FAM-VAD-FMK (*see* **Subheading 3.3.2.**, step 3). There is a low level of spontaneous apoptosis as would be expected in the fast-growing SCK tumor. The second photograph shows the ATO-treated test mouse that received an IV injection of FAM-VAD-FMK (*see* **Subheading 3.3.2.**, step 3). There is a high level of apoptosis as a result of treatment with ATO (*see* **Subheading 3.3.2.**, step 2) after 6 h.

17. To prevent photo-bleaching of fluorescent-stained apoptotic cells, minimize exposure to light as much as possible during this procedure.

18. The results of the flow cytometry analysis demonstrate that ATO induces a greater level of apoptosis in SCK tumor cells (39% in ATO-treated mouse) when compared with only 18% apoptosis induction in the saline control mouse. Cell permeant fluorescent-labeled FMK peptides that bind to active caspases can be used to identify areas of apoptosis in living animals by IV injection. The cell membrane permeability allows the probes to enter into several layers of tumor cells and tissues. Unbound probe is naturally removed from non-apoptotic cells in the living

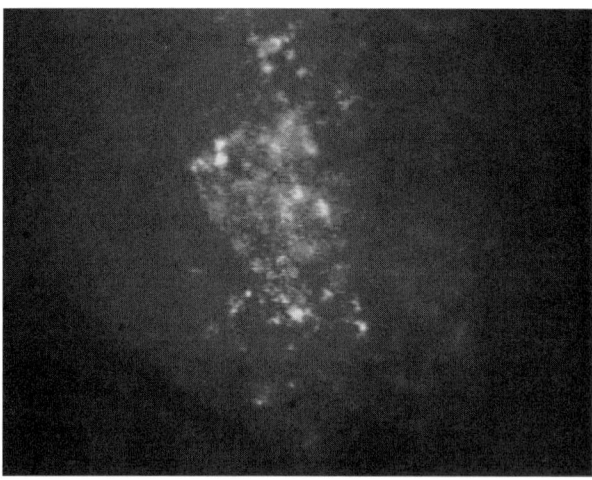

Fig. 11. *In vivo* apoptosis detection in an ATO-treated mouse, 7-day-old SCK tumor. The test mouse received an ATO (8 mg/kg IP injection). The test mouse subsequently received an IV injection of 8.0 μg cell permeant FAM-VAD-FMK poly-caspase detection probe 6 h after ATO injection. The probe was allowed to circulate for 30 min before a 20× fluorescent image was captured using a Hamamatsu C2400 camera and Broadway Imaging Software on an Eclipse TE200 bench-top microscope.

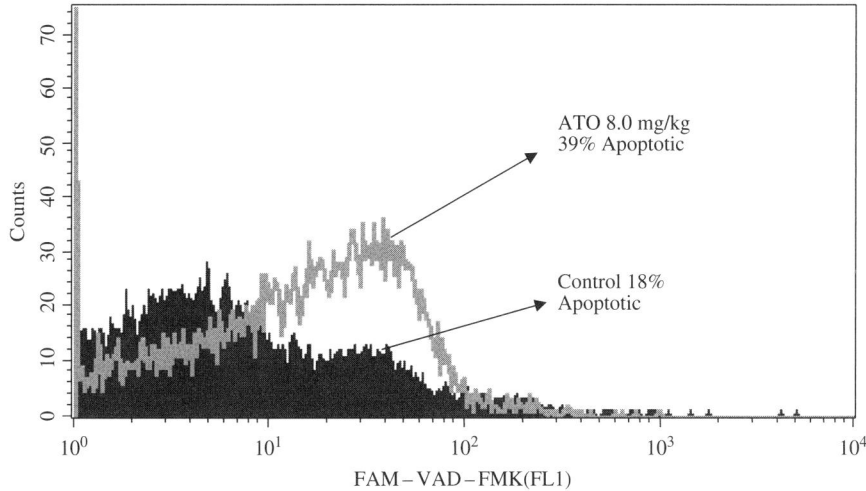

Fig. 12. Flow cytometry histogram overlay of FAM-VAD-FMK-treated tumor cells taken from either the saline-treated or the ATO-treated tumor masses (unpublished data). In these results, the tumors were scraped from the window chambers 30 min after IV injection with FAM-VAD-FMK. The tumors were excised, and cell suspensions were made by dispersing the cells with trypsin. Cell suspensions were then analyzed by flow cytometry. The increase in overall fluorescence (right shift) of the ATO-treated tumor mice (cells) relative to that observed in the saline control mice reflects the increased apoptosis induction rate generated by the ATO exposure (*see* **Note 19**).

    animal, and bound probe remains within the apoptotic cells. This is an effective method of detecting and measuring apoptosis *in vivo*.

19. The flow cytometry results support the results seen in the photographs (*see* **Figs 10 and 11**)is a detectable level of apoptosis in the control mice (placebo-treated) tumors and a significant increase in apoptosis in the test mice (ATO-treated) tumors.

References

1. Kerr, J.F.R., Wyllie, A.H., and Currie, A.R. (1972) Apoptosis: a basic biological phenomenon with wide-ranging implications in tissue kinetics. *Br. J. Cancer* **26**, 239–257.
2. Wyllie, A.H., Kerr, J.F.R., and Currie, A.R. (1981) Cell death: the significance of apoptosis. *Int. Rev. Cytol.* **68**, 251–306.
3. Earnshaw, W.C., Martins, L.M., and Kaufmann, S.H. (1999) Mammalian caspases: structure, activation, substrates, and functions during apoptosis. *Annu. Rev. Biochem.* **68**, 383–424.

4. Nicholson, D.W. (1999) Caspase structure, proteolytic substrates, and function during apoptotic cell death. *Cell Death Differ.* **6**, 1028–1042.

5. Heemels, M.T. (ed.) (2000) Nature insight apoptosis. *Nature* **407**, 770–816.

6. Degterev, A., Boyce, M., and Yuan, J. (2003) A decade of caspases. *Oncogene* **22**, 8543–8567.

7. Fink, S.L. and Cookson, B.T. (2005) Apoptosis, pyroptosis, and necrosis: mechanistic description of dead and dying eukaryotic cells. *Infect. Immun.* **73**, 1907–1916.

8. Lavrik, I.N., Golks, A., and Krammer, P.H. (2005) Caspases: pharmacological manipulation of cell death. *J. Clin. Invest.* **115**, 2665–2672.

9. Thornberry, N.A. and Lazebnik, Y. (1998) Caspases: enemies within. *Science* **281**, 1312–1316.

10. Thornberry, N.A. et al. (1992) A novel heterodimeric cysteine protease is required for interleukin-1β processing in monocytes. *Nature* **356**, 768–774.

11. Thornberry, N.A. et al. (1997) A combinatorial approach defines specificities of members of the caspase family and granzyme B. *J. Biol. Chem.* **272**, 17907–17911.

12. Fuentes-Prior, P. and Salvesen, G.S. (2004) The protein structures that shape caspase activity, specificity, activation and inhibition. *Biochem. J.* **384**, 201–232.

13. Sarin, A., Wu, M.L., and Henkart, P.A. (1996) Different interleukin-1β converting enzyme (ICE) family protease requirements for the apoptotic death of T lymphocytes triggered by diverse stimuli. *J. Exp. Med.* **184**, 2445–2450.

14. Armstrong, R.C. et al. (1996) Fas-induced activation of the cell death-related protease CPP32 is inhibited by Bcl-2 and by ICE family protease inhibitors. *J. Biol. Chem.* **271**, 16850–16855.

15. Garcia-Calvo, M., Peterson, E.P., Leiting, B., Ruel, R., Nicholson, D.W., and Thornberry, N.A. (1998) Inhibition of human caspases by peptide-based macromolecular inhibitors. *J. Biol. Chem.* **273**, 32608–32613.

16. Sun, X.M., MacFarlane, M., Zhuang, J., Wolf, B.B., Green, D.R., and Cohen, G.M. (1999) Distinct caspase cascades are initiated in receptor-mediated and chemical-induced apoptosis. *J. Biol. Chem.* **274**, 5053–5060.

17. Ekert, P.G., Silke, J., and Vaux, D.L. (1999) Caspase inhibitors. *Cell Death Differ.* **6**, 1081–1086.

18. Clark, P., Dziarmaga, A., Eccles, M., and Goodyer, P. (2004) Rescue of defective branching nephrogenesis in renal coloboma syndrome by the caspase inhibitor, z-VAD-FMK. *J. Am. Soc. Nephrol.* **15**, 299–305.

19. Nakano, M. et al. (2004) Caspase-3 inhibitor prevents apoptosis of human islets immediately after isolation and improves islet graft function. *Pancreas* **29**, 104–109.

20. Nedev, H.N., Klaiman, G., LeBlanc, A., and Saragovi, H.U. (2005) Synthesis and evaluation of novel dipeptidyl benzoyloxymethyl ketones as caspase inhibitors. *Biochem. Biophys. Res. Commun.* **336**, 397–400.

21. Rauber, P., Angliker, H., Walker, B., and Shaw, E. (1986) The synthesis of peptidylfluoromethanes and their properties as inhibitors of serine proteases and cysteine proteases. *Biochem. J.* **239**, 633–640.
22. Powers, J.C., Asgian, J.L., Ekici, O.D., and James, K.E. (2002) Irreversible inhibitors of serine, cysteine, and threonine proteases. *Chem. Rev.* **102**, 4639–4750.
23. Wu, J.C. and Fritz, L.C. (1999) Irreversible caspase inhibitors: tools for studying apoptosis. *Methods* **17**, 320–328.
24. Bedner, E., Smolewski, P., Amstad, P., and Darzynkiewicz, Z. (2000) Activation of caspases measured *in situ* by binding of fluorochrome-labeled inhibitors of caspases (FLICA): correlation with DNA fragmentation. *Exp. Cell Res.* **259**, 308–313.
25. Amstad, P.A., Yu, G.L., Johnson, G.L., Lee, B.L., Dhawan, S., and Phelps, D.J. (2001) Detection of caspase activation *in situ* by fluorochrome-labeled caspase inhibitors. *Biotechniques* **31**, 608–616.
26. Smolewski, P., Bedner, E., Du, L., Hsieh, T.C., Wu, J.M., Phelps, D.J., and Darzynkiewicz, Z. (2001) Detection of caspase activation by fluorochrome-labeled inhibitors: multiparameter analysis by laser scanning cytometry. *Cytometry* **44**, 73–82.
27. Smolewski, P., Grabarek, J., Halicka, H.D., and Darzynkiewicz, Z. (2002) Assay of caspase activation *in situ* combined with probing plasma membrane integrity to detect three distinct stages of apoptosis. *J. Immunol. Methods* **265**, 111–121.
28. Smolewski, P., Grabarek, J., Lee, B.W., Johnson, G.L., and Darzynkiewicz, Z. (2002) Kinetics of HL-60 cell entry to apoptosis during treatment with TNF-a or camptothecin assayed by the stathmo-apoptosis method. *Cytometry* **47**, 143–149.
29. Wolbers, F., Buijtenhuijs, P., Haanen, C., and Vermes, I. (2004) Apoptotic cell death kinetics *in vitro* depend on the cell types and the inducers used. *Apoptosis* **9**, 385–392.
30. Grunewald, S., Paasch, U., Said, T.M., Sharma, R.K., Glander, H.J., and Agarwal, A. (2004) Caspase activation in human spermatozoa in response to physiological and pathological stimuli. *Fertil. Steril.* **83**, 1106–1112.
31. Fiala, M., Lin, J., Ringman, J., Arab, V.K., Tsao, A., Patel, A., Lossinsky, A.S., Graves, M.C., Gustavson, A., Sayre, J., Sofroni, E., Suarez, T., Chiappelli, F., and Bernard, G. (2005) Ineffective phagocytosis of amyloid-B by macrophages of Alzheimer's disease patients. *J. Alzheimer Dis.* **7**, 221–232.
32. Mizukami, S., Kikuchi, K., Higuchi, T., Urano, Y., Mashima, T., Tsuruo, T., and Nagano, T. (1999) Imaging of caspase-3 activation in HeLa cells stimulated with etoposide using a novel fluorescent probe. *FEBS Lett.* **453**, 356–360.
33. Hug, H., Los, M., Hirt, W., and Debatin, K.M. (1999) Rhodamine 110-linked amino acids and peptides as substrates to measure caspase activity upon apoptosis induction in intact cells. *Biochemistry* **38**, 13906–13911.
34. Mack, A., Furmann, C., and Hacker, G. (2000) Detection of caspase-activation in intact lymphoid cells using standard caspase substrates and inhibitors. *J. Immunol. Methods* **241**, 19–31.

35. Komoriya, A., Packard, B.Z., Brown, M.J., Wu, M.L., and Henkart, P.A. (2000) Assessment of caspase activities in intact apoptotic thymocytes using cell-permeable fluorogenic substrates. *J. Exp. Med.* **191**, 1819–1828.

36. Telford, W.G., Komoriya, A., and Packard, B.Z. (2002) Detection of localized caspase activity in early apoptotic cells by laser scanning cytometry. *Cytometry* **47**, 81–88.

37. Lee, B.W., Johnson, G.L., Hed, S.A., Darzynkiewicz, Z., Talhouk, J.W., and Mehrotra, S. (2003) DEVDase detection in intact apoptotic cells using the cell permeant fluorogenic substrate, (z-DEVD)2-cresyl violet. *Biotechniques* **35**, 1080–1085.

38. Li, X., Du, L., and Darzynkiewicz, Z. (2000) During apoptosis of HL-60 and U-937 cells caspases are activated independently of dissipation of mitochondrial electrochemical potential. *Exp. Cell Res.* **257**, 290–297.

39. Duan, R.W., Garner, D.S., Williams, S.D., Funckes-Shippy, C.L., Spath, I.S., and Blomme, E.A. (2003) Comparison of immunohistochemistry for activated caspase-3 and cleaved cytokeratin 18 with TUNEL method for quantification of apoptosis in histological sections of PC-3 subcutaneous xenografts. *J. Pathol.* **199**, 221–228.

40. Salvioli, S., Ardizzoni, A., Franceschi, C., and Cossarizza, A. (1997) JC-1 but not DiOC6(3) or rhodamine 123, is a reliable fluorescent probe to assess delta psi changes in intact cells: implications for studies on mitochondrial functionality during apoptosis. *FEBS Lett.* **411**, 77–82.

41. Poot, M., Zhang, Y.Z., Kramer, J.A., Wells, K.S., Jones, L.J., Hanzel, D.K., Lugade, A.G., Singer, V.L., and Haugland, R.P. (1996) Analysis of mitochondrial morphology and function with novel fixable fluorescent stains. *J. Histochem. Cytochem.* **44**, 1363–1372.

42. Brunner, K.T., Mauel, J., Cerottini, J.C., and Chapuis, B. (1968) Quantitative assay of the lytic action of immune lymphoid cells on 51-Cr-labeled allogenic target cells *in vitro*; inhibition by isoantibody and by drugs. *Immunology* **14**, 181–196.

43. Ward, P.W., Bonaparte, M.I., and Barker, E. (2004) HLA-C and HLA-E reduce antibody-dependent natural killer cell-mediated cytotoxicity of HIV-infected primary T cell blasts. *AIDS* **18**, 1769–1779.

44. Barber, D.L., Wherry, E.J., and Ahmed, R. (2003) Cutting edge: rapid *in vivo* killing by memory CD8 T cells. *J. Immunol.* **171**, 27–31.

45. Yang, S. and Haluska, F.G. (2004) Treatment of melanoma with 5-fluorouracil or dacarbazine *in vitro* sensitizes cells to antigen-specific CTL lysis through perforin/granzyme- and Fas-mediated pathways. *J. Immunol.* **172**, 4599–4608.

46. Schwarer, A.P., Jiang, Y.Z., Deacock, S., Brookes, P.A., Barret, A.J., Goldman, J.M., Batchlor, J.R., and Lechler, R.I. (1994) Comparison of helper and cytotoxic antirecipient T cell frequencies in unrelated bone marrow transplantation. *Transplantation* **58**, 1198–1203.

47. Russell, J.H. and Ley, T.J. (2002) Lymphocyte-mediated cytotoxicity. *Annu. Rev. Immunol.* **20**, 323–370.

48. Pross, H., Callewaert, D., and Rubin, P. (1986) Assays for NK cell cytotoxicity-their values and pitfalls, in *Immunobiology of Natural Killer Cells* (E. Lotzova and R.B. Herberman eds.) Vol. I. CRC Press, Boca Raton, FL, pp. 1–16.

49. Jerome, K.R., Sloan, D.D., and Aubert, M. (2003) Measurement of CTL-induced cytotoxicity: the caspase 3 assay. *Apoptosis* **8**, 563–571.

50. Slezak, S.E. and Horan, P.K. (1989) Cell-mediated cytotoxicity: a highly sensitive and informative flow cytometric assay. *J. Immunol. Methods* **117**, 205–214.

51. Radosevic, K., Garritsen, H.S.P., Van Graft, M., De Grooth, B.G., and Greve, J. (1990) A simple and sensitive flow cytometric assay for the determination of the cytotoxic activity of human natural killer cells. *J. Immunol. Methods* **135**, 81–89.

52. Hatam, L., Schuval, S., and Bonagura, V.R. (1994) Flow cytometric analysis of natural killer cell function as a clinical assay. *Cytometry* **16**, 59–68.

53. Lee-MacAry, A.E., Ross, E.L., Davies, D., Laylor, R., Honeychurch, J., Glennie, M.J., Snary, D., and Wilkinson, R.W. (2001) Development of a novel flow cytometric cell-mediated cytotoxicity assay using the fluorophores PKH-26 and TO-PRO-3 iodide. *J. Immunol. Methods* **252**, 83–92.

54. Hoppner, M., Luhm, J., Schlenke, P., Koritke, P., and Frohn, C. (2002) A flow-cytometry based cytotoxicity assay using stained effector cells in combination with native target cells. *J. Immunol. Methods* **267**, 157–163.

55. Olin, M.R., Choi, K.H., Lee, J., and Molitor, T.W. (2005) $\gamma\delta$ T-lymphocyte cytotoxic activity against *Mycobacterium bovis* analyzed by flow cytometry. *J. Immunol. Methods* **297**, 1–11.

56. Kienzle, N., Oliver, S., Buttigieg, K., and Kelso, A. (2002) The fluorolysis assay, a highly sensitive method for measuring the cytolytic activity of T cells at very low numbers. *J. Immunol. Methods* **267**, 98–108.

11

Caspase Activity Assays

Andrew L. Niles, Richard A. Moravec, and Terry L. Riss

Summary

Caspase activity assays in multi-well plate formats represent powerful tools for understanding experimental modulation of the apoptotic response. These assays are configured to exploit functional, biochemical, and temporal differences in substrate specificity and selectivity, which are useful in defining the magnitude and mechanism of a compound or treatment effect. New advances in fluorescent and luminescent chemistries now enable single addition "add-mix-measure" determinations of caspase activity directly in the sample plate with unprecedented sensitivity. Unlike other more cumbersome or laborious techniques, caspase activity induction or inhibition measures are quantifiable and definitive. The highlighted techniques in this chapter are cost efficient and allow for the rapid exploration of thousands of combinations and conditions.

Key Words: Caspase; activity; selectivity; cell based; induction; inhibition; fluorescence; luminescence.

1. Introduction

Caspase activation is a central biochemical process in apoptosis. The proteolytic activities of the caspases are known to initiate and augment the apoptotic response by maintenance of the enzymatic cascade in a committed program. Furthermore, activated caspases degrade both important structural protein elements within cells and RNA splicing or DNA repair-associated proteins (*1–3*). Ultimately, this group of enzymatic activities is responsible for dramatic phenotypic changes and elimination of the apoptotic cell. The caspase family of proteins can be conveniently distinguished from other cellular proteolytic activities by their activity profile. In the most general sense, this group

From: *Methods in Molecular Biology, vol. 414: Apoptosis and Cancer*
Edited by: G. Mor and A. B. Alvero © Humana Press Inc., Totowa, NJ

of proteases utilizing a cysteine residue in their active site, has an absolute requirement for substrates containing aspartic acid residues *(4–6)*. In fact, these shared attributes were the genesis of the term "caspase" (for cysteinyl aspartate-specific proteinase) *(7)*.

1.1. The Biochemical Basis for Substrate Selectivity

The activity profiles and substrate preference of particular caspases have been previously defined by Thornberry et al. *(8)*. Using recombinantly expressed caspases 1, 2, 3, 4, 5, 7, 8, and 9, and a combinatorial substrate approach varying P4, P3, and P2 residues with respect to the scissile cleavage site (Schechter–Berger nomenclature), they found that these enzymes could be grouped into three major groups by substrate specificity (*see* Table 1). For each group, the primary basis for substrate preference can be attributed to the influence of the P4 amino acid.

Group I caspases prefer the tetrapeptide sequence WEHD but are poorly selective. Caspases 1, 4, and 5, however, have an ill-defined role in apoptosis and are likely to participate in other biological cascades such as the exacerbation of inflammatory responses modulated by interleukin-β and interferon-γ-inducing factor *(9,10)*.

Group II caspases have a striking preference for the tetrapeptidic sequence of DE(X)D, whereby the P2 amino acid residue dictates a high degree of selectivity. Caspase-2 demonstrates a further preference for pentapeptide substrates and favors valine at P5 position *(11)*. In addition to a substantially restricted substrate usage pattern, caspases 3 and 7 demonstrate unparalleled catalytic

Table 1
The caspase family and consensus peptide useage.

Family grouping	Caspase	Optimal sequence (P4-P3-P2-P1)
Group I	Caspase-1	WEHD
	Caspase-4	(W/L)EHD
	Caspase-5	(W/L)EHD
Group II	Caspase-3	DEVD
	Caspase-7	DEVD
	Caspase-2	VD(V/E)(A/H)D
Group III	Caspase-6	VEHD
	Caspase-8	LETD
	Caspase-9	LEHD

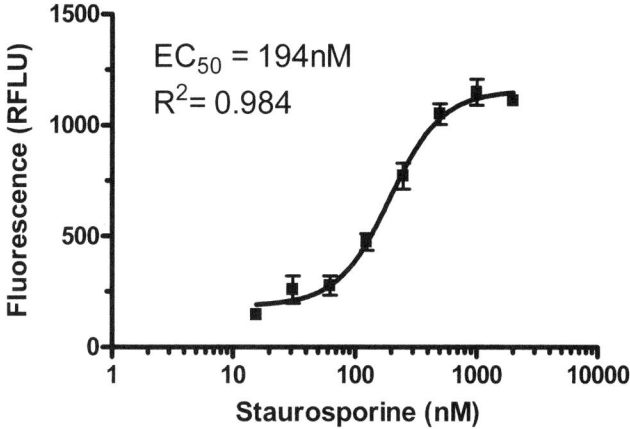

Fig. 1. Dose-response curve of caspase-3/7 activation in staurosporine-treated Jurkat cells. Jurkat cells were plated at 10,000 cells per well in 50 μl volumes in RPMI 1640 + 10% FBS. Serial dilutions of staurosporine (Biomol) or DMSO (Sigma) matched vehicle were added to cells in an additional 50 μl volume. After 5 h of incubation at 37°C, the plate was removed from the incubator and allowed to equilibrate to room temperature. Apo-ONE® Reagent (Promega) was added in an additional volume and briefly mixed by orbital shaking. After 1 h, the resulting fluorescence was measured using a CytoFluor II. The data were then plotted and fit by GraphPad Prism™ software. EC_{50} value represents half-maximal caspase-3/7 activation. The R^2 value represents the goodness of fit. A.L Niles, unpublished data.

efficiencies against their favored substrates when compared with members of the other groups *(12)*. Furthermore, group II caspases demonstrate a high level of intrinsic activity repression known as "zymogenicity," when maintained in an inactivated, proenzyme form *(13,14)*.

The group III caspases utilize the general consensus sequence (L/V)E(X)D. Selectivity among caspases 6, 8, and 9 is imparted by the P4 and P2 residues; especially for caspase-9 that has a heavy preference for a histidine residue at P2. The group III caspases have a selectivity profile akin to group I caspases and are more promiscuous than group II caspases. This grouping also demonstrates a higher level of proenzyme activity than the group II enzymes because their activities are less tightly controlled by proenzyme leader sequences.

1.2. The Functional Basis of Differential Activity

The caspases directly involved in apoptosis can also be sub-grouped by their functional role in the process. Although the initiator caspases 8 and 9

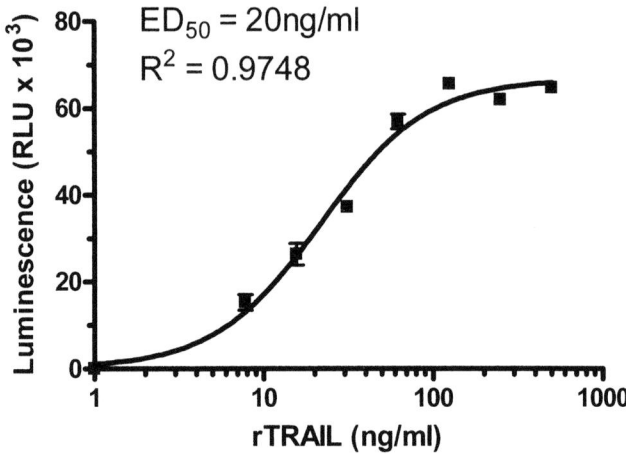

Fig. 2. Dose-response curve of caspase-8 activation in recombinant TRAIL-treated Jurkat cells. Jurkat cells were plated at 25,000 cells per well in 50 μl volumes in RPMI 1640 + 10% FBS. Serial dilutions of rTRAIL (Chemicon) or matched vehicle were added to cells in an additional 50 μl volume. After 5 h of incubation at 37°C, the plate was removed from the incubator and allowed to equilibrate to room temperature. Caspase-Glo®-8 Reagent (Promega) was added in an additional volume and briefly mixed by orbital shaking. After 1 h, the resulting luminescence was measured using a BMG PolarStar™. The data were then plotted and fit by GraphPad Prism™ software. The EC_{50} value represents half-maximal caspase-8 activation. The R^2 value represents the goodness of fit. A.L Niles, unpublished data.

both contain long pro-domains, it is well appreciated that each participates in distinctly different initiation pathways. Caspase-8 activity is typically triggered by a sequence of events beginning with cell surface receptor engagement of the TNF superfamily by agonistic ligands (15,16). The binding of these ligands allows for the assembly of a complex of proteins and adapter molecules that lead to autocatalytic processing of pro-caspase-8 into a mature and fully active enzyme. This activation mechanism is often referred to as the "extrinsic" pathway.

The activation of caspase-9 by the "intrinsic" pathway involves cell death signals derived from mitochondria. The stimulus for this event is perceived damage to the cell from environmental stresses or accumulated metabolic toxins. In this activation event, released cytochrome C complexes with Apaf-1 and pro-caspase-9 in the presence of ATP to form a complex called the apoptosome. The complex then becomes catalytically competent (17).

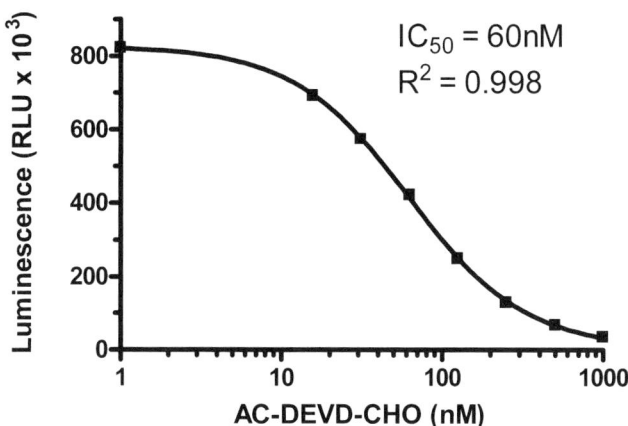

Fig. 3. Caspase-8 activity inhibition by Ac-DEVD-CHO. Ac-DEVD-CHO (Promega) was twofold serially diluted in RPMI 1640. DMSO matched vehicle served as controls. Recombinant human caspase-8 (Biomol) was diluted to 100 U/ml in RPMI 1640, then added in an equal volume to wells containing caspase inhibitor or vehicle controls. After 30 min of incubation at room temperature, Caspase-Glo®-8 Reagent (Promega) was added in an additional 100 μl volume. Luminescence was measured 60 min after addition using a BMG FluoStar™. The data were plotted and fit by GraphPad Prism™ software. IC_{50} value refers to the half-maximal inhibition of caspase-8 activity by Ac-DEVD-CHO. The R^2 value represents the goodness of fit. A.L Niles, unpublished data.

Caspases 3 and 7 are distinct from the initiator caspases in that they mediate effector or "executioner" functions downstream from the original stimulus. In fact, caspases 3 and 7 require processing of their proenzyme leader sequence by caspase-8 and/or caspase-9 to become activated. Independent of specific induction pathway, the apoptotic program convergence at caspases 3 and 7 allows for a critical checkpoint in commitment for cellular regulation.

The precise role of caspases 2 and 6 remain to be fully elucidated *(18–20)*. Because pro-caspase-2 contains the caspase recruitment domain (CARD), it is typically grouped as an initiator caspase. Caspase-6 appears to have a clearer role as an effector caspase. Nevertheless, there is substantial evidence to suggest both enzymes contribute to feedback amplification loops in the apoptotic cascade *(21)*.

As the study of apoptosis has progressed, so too have the tools and assays available to measure the process. Early morphological observations of cells undergoing apoptosis gave way to more specific and direct techniques such as

TUNEL and annexin-V labeling of cells. Improvements in antibody availability enabled western blot and immunohistochemical analyses of cleaved proteins such as PARP and caspase-3. New techniques and chemistries have evolved that further simplify caspase assays and allow multi-well plate-based analyses, and these will be the scope of this chapter.

2. Materials
2.1. Equipment

1. 15-ml conical tubes (Corning, Lowell, MA, USA).
2. 75 cm^2 culture flasks (Corning).
3. Reagent reservoirs (Corning).
4. Sterile pipette tips.
5. Single and multi-channel micro-pipettors.
6. Hemocytometer and coverslip.
7. Bright-field microscope.
8. Table-top centrifuge.
9. Opaque-walled 96-well plate (Corning).
10. Fluorometer equipped with $485_{ex} \cdot 527_{em}$ optical filters.
11. Luminometer.
12. CO2 incubator.

2.2. Reagents

1. Cell-culture medium (RPMI, MEM, DMEM, etc.).
2. Animal serum (fetal bovine, horse, etc.).
3. Trypan blue solution (Sigma, St. Louis, MO, USA).
4. Staurosporine (Biomol, Plymouth Meeting, PA, USA).
5. Dimethyl sulfoxide (DMSO) (Sigma).
6. Recombinant TRAIL (Chemicon, Temecula, CA, USA).
7. Recombinant caspase-8 (Biomol).
8. HEPES hemisodium salt (Sigma).
9. Dithiothreitol (DTT) (Sigma).
10. Ac-DEVD-CHO caspase inhibitor (Promega, Madison, WI, USA).
11. Apo-ONE® Homogeneous Caspase-3/7 Assay (Promega).
12. Caspase-Glo®-3/7, 8, and 9 Activity Assays (Promega).

3. Methods

The methods described below outline experimental (i) caspase induction in cell culture, (ii) biochemical caspase inhibition, (iii) fluorogenic caspase activity measurement, and (iv) luminogenic caspase activity measurement.

3.1. Caspase Activation

Cell-culture manipulation should be conducted in a laminar flow hood or clean-room environment with sterile reagents and consumables. Aseptic technique should be employed to avoid culture infection.

1. Harvest cells from culture. Remove a representative volume (1000 μl or less) for analysis. Gently pellet the remaining pool by centrifugation at $200 \times g$ for 8–10 minutes at room temperature.
2. Examine a dilution of the cell sample by trypan blue exclusion to determine the population viability and cell count. If viability is greater than 90%, add complete medium (with serum and nutrient/cofactor adjuncts) so that the cells are at a density of 200,000 viable cells/ml.
3. Plate the cells in 50 μl volumes (10,000cells/well) to a sterile, opaque-walled 96-well plate. Add 100 μl volumes of cell-culture medium to replicate wells to serve as cell-free control. If the cells are attachment dependent, allow for a 2–16 h equilibration period before proceeding.
4. (i) Prepare caspase induction control agents. For example, dilute staurosporine to 4 μM (or rTRAIL to 500 ng/ml) in complete cell-culture medium and conduct twofold serial dilutions. Add the solutions in replicate 50 μl volumes to cell wells for a final concentration range between 2 μM and 15.6 nM. (ii) Dilute unknown or test compound to 20 μM in complete medium and add them in 50 μl volumes to cell wells. (iii) Add 50 μl complete cell-culture medium to wells designated for non-induction, negative control. Mix the plate by orbital shaking to insure homogeneity and compound dispersion.
5. Incubate the plate at 37°C in a humidified CO2 environment for period between 4 and 24 h (determined empirically).
6. Proceed to activity assay.

3.2. Caspase Inhibition

Inhibition studies are useful for identifying and characterizing the agents that negatively modulate caspase activity. When planning the experiment, tailor the assay conditions to reflect the biology of the particular caspase enzyme of interest. Although most secondary potency testing is conducted with purified caspases in defined biochemical assay systems, more complex inhibition assays can be performed with cell-culture models.

1. (i) Dilute test compounds to 20 μM in an appropriate assay buffer. In most cases, a buffer constructed of 50 mM HEPES, pH 7.2, with 5–10 mM DTT will suffice. For example, make an initial dilution of Ac-DEVD-CHO to 20 μM twofold serially and dilute the test compounds in a micro-well plate in 50 μl volumes with a multi-channel pipettor. Replicate wells should be created for each concentration of

compound so that wells with enzyme and without enzyme are represented. (ii) Add 50 µl volumes of assay buffer to replicate wells to serve as uninhibited control.

2. Dilute the target caspase in the above assay buffer to a concentration that is known to be in the middle of the activity linear range of the detection assay to be used. This must be determined empirically prior to inhibition studies.

3. (i) Add the diluted caspase to replicate wells containing dilutions of inhibitor and control wells not containing inhibitor. (ii) Add assay buffer only to parallel background control wells containing inhibitor dilutions. Mix by orbital shaking and incubate at 37°C until inhibition equilibrium; a period usually between 30 min and 1 h.

4. Proceed to activity assay.

3.3. Fluorescent Caspase Activity Assays

The proteolytic activity of activated caspases can be measured in multi-well formats by using profluorescent tetrapeptide substrates presented in a buffer. In the absence of caspase activity, profluorescent substrates demonstrate poor excitation and emission efficiency because of a peptidic quenching effect. After caspase cleavage at the scissile bond linkage, the fluorophore is liberated from the peptide. This free fluorophore exhibits a substantial increase in quantum yield under optimal excitation. Because fluorophore accumulation is proportional to caspase activity, the resulting fluorescence can be measured using a fluorometer in either kinetic or endpoint experiments.

The homogeneous assay format offers a more convenient and robust protocol than conventional multi-step methods by eliminating the need for wash and centrifugation steps. In this protocol, a bifunctional buffer system mediates lysis and supports optimal caspase activity in a single add and mix procedure. This allows for interrogation of caspase activity directly in cell-culture medium. Rhodamine 110 (R110)-derived peptide substrates are typically utilized in commercial kits because this fluorophore permits greater assay sensitivity than coumarin-derivatized substrates in 96 and 384 multi-well formats (22).

The following is an abbreviated protocol for performing a homogeneous assay to measure caspase-3/7 activity from cells in culture. For complete details of the assay format and potential usages, consult Promega Technical Bulletin #295 that describes the Apo-ONE® Homogeneous Caspase-3/7 Assay.

1. Thaw assay buffer and 100 × substrate and allow them to equilibrate to room temperature. Mix by manual inversion or by vortexing.

2. Dilute substrate 1:100 with the buffer to obtain the desired volume of Apo-ONE® Homogeneous Caspase-3/7 Reagent.

3. Add the Apo-ONE® Reagent to assay wells in a 1:1 ratio (e.g., 100 µl reagent to 100 µl sample or control).

4. Mix to homogeneity by gentle orbital shaking. Abstain from manual aspiration and dispensing to avoid introduction of bubbles that complicate fluorescence measurements.
5. Incubate the sample/reagent mixes shielded from ambient light for 1–18 h prior to measuring the fluorescence using a fluorometer. The excitation and emission optima for R110 are 499 and 521 nm, respectively. Paired filter sets in the range of 485 ± 20 and 530 ± 20 nm are sufficient to accurately detect the Stoke's shift associated with the profluorescent and free fluorescent product.

3.4. Luminescent Caspase Activity Assays

Luminescent caspase activity assays are also configured in a homogeneous format and offer an alternative to fluorescent assays *(23)*. These homogeneous assays utilize amino-luciferin-derivatized peptide substrates configured in a buffer system containing luciferase, MgSO4, and ATP. In the absence of caspase activity, the amino-luciferin-derivatized peptide substrate is an exceedingly poor substrate for luciferase and as such emits minimal light. In the presence of caspase activity, however, the peptide moiety is removed from the substrate, leaving amino-luciferin that is rapidly consumed by luciferase to generate light that can be measured using a luminometer. This dual enzyme system demonstrates a proportional response between caspase activity and resulting stably sustained photon emission, known as a "glow-type luminescence."

The following is an abbreviated protocol for performing a homogeneous luminescent assay to measure caspase activity from cells in culture. For complete details of the assay format and potential usages, consult Promega Technical Bulletins #323, 332, and 333 that describe the Caspase-Glo®-3/7, -8 and -9 Assays.

1. Equilibrate the frozen Caspase-Glo® buffer and lyophilized Caspase-Glo® substrate to room temperature prior to use.
2. Transfer the Caspase-Glo® buffer to the lyophilized Caspase-Glo® substrate to create the reagent. Mix by swirling or by inverting the contents until the substrate is completely dissolved.
3. Add an equal volume of Caspase-Glo® reagent to sample wells containing cells, enzyme, or controls, and mix well by orbital shaking. Do not mix the sample/reagent mix by manual aspiration and dispensing as this may introduce bubbles that complicate luminescent measurements.
4. Incubate the samples at room temperature for 15–30 min.
5. Measure luminescence on a luminometer. Actual incubation times (step 4) should be determined empirically by repeated measurements in kinetic mode. The signal derived at peak luminescence should be stable for at least 2 h after reagent addition.

4. Notes

4.1. Cell-Culture Biology Considerations

Cell-based models of caspase induction or inhibition can be extremely effective and efficient tools for dissecting apoptosis pathways or evaluating potential therapeutic compound efficacy. It should be noted, however, that the quality of the data derived from these studies is substantially dependent on careful attention to experimental design and detail. In short, understanding the test in vitro cell model and its limitations strengthens the data and ultimate conclusion *(24)*.

1. An early and important aspect of understanding the cell-culture model is to establish cell growth parameters for normal physiology or proliferation during the experiment. Normal doubling times and cellular volume vary greatly and impact initial seeding density. In general, a density should be chosen in which cells will not become contact inhibited, form multiple attachment layers, or require in-experiment replenishment of medium or growth factors. Typical 96-multi-well formats accommodate anywhere from 2500 to 10,000 cells/well.

2. Values derived from cell-based caspase activity assays represent an average response to the treatment. Because of heterogeneity in the population of cells within the well and position in the cell cycle, subsets of the population may be more or less susceptible to caspase induction. Although cell cycle-synchronizing reagents exist, their use is strongly discouraged because of unnecessary system perturbation and possible unintended effects.

3. The kinetics of the caspase activation response should be examined with respect to compound or treatment dosages and exposures. Several different concentrations of a candidate compound should be added to cells in multiple wells while effects are observed over the contact time. This approach not only helps to reveal the optimal activation potency of particular subsets of caspase enzymes but addresses the temporal aspects of that activation. For instance, the initiator caspases are activated early in the cascade then decline in activity in accordance with their enzymatic half-life. Conversely, effector caspase activities are less proximal to the stimulus event and are largely maintained until cells proceed into secondary necrosis. Any kinetic examination of caspase activation should include well-matched vehicle controls or mock treatments to identify the magnitude of the specific effect. Similarly, known inducing agents are useful to demonstrate the model system caspase activation potential.

4. Inhibition experiments can be used to dissect critical checkpoints in the caspase cascade or to identify and characterize the potency of potentially therapeutic compounds or treatments. Inhibition studies in cell-based systems provide a wealth of information in regards to compound stability and permeability within the context of a complex biological environment. Data from cell-based assays provide a starting point for ascertaining the practical potency of the inhibitor and potential biophysical

limitations. For instance, small molecule inhibitors may exhibit excessive serum-binding kinetics that substantially limit the free molecule concentrations that can be obtained.

However, a common pitfall of cell-based, multi-well caspase inhibition studies is the initiation of non-caspase cytotoxicity by the experimental treatment. Erroneous conclusions therefore may be made with regard to inhibition effects or potency. This problem can be mitigated by normalizing assay readouts for cytotoxicity or viability by adding assay reagents to confirm viability status. The specificity of the inhibitor response is also critical in this regard, as caspase pathway inhibitors might have broad spectrum effects compromising cellular health. Inhibitor specificity therefore should be addressed with not only purified target but also structurally or functionally related caspase enzyme panels prior to cell-based studies, when feasible.

Lastly, cell-based inhibition studies using exogeneously added chemical mediators should include experimental design measures to remove non-internalized and bound inhibitor. This can be accomplished by a wash step prior to assaying for caspase activity. This step is necessary because lytic endpoint caspase activity assays measure remaining activity in a lysate pool. Therefore, unbound inhibitor may access and inhibit previously uninhibited caspases liberated by the assay lysis reagent. The obvious caveat to washing the treated cell populations is that profoundly apoptotic cells may have progressed to necrosis and released their activated caspases. Similarly, cells undergoing apoptosis often "lift off" culture wells during washes. These losses may lead to underestimation of the caspase activation response in the uninhibited control wells.

4.2. Fluorescent Caspase Activity Assay Considerations

1. Not all fluorescent activity assays are well suited for cell-based models. Enzymes that demonstrate poor catalytic efficiency or are non-abundant suffer from low sensitivity.
2. The largest detriment to fluorescent caspase activity assays is interference by the sample compound with the fluorescence measurement of the reporting molecule. For instance, color quenching by candidate compounds or bubbles in the sample can impact the efficiency of excitation or emission. Similarly, auto-fluorescent compounds may contribute fluorescence to the emission channel in the absence of caspase activity and accumulated fluor (false positive).
3. Activity assay incubation times must be empirically determined to achieve the highest signal to background ratio. Repeated measurements are not detrimental to fluorescent product accumulated as a result of caspase cleavage of the substrate.

4.3. Luminescent Activity Assay Considerations

1. Luminescent signal output is affected by temperature. To mitigate variation in assay signal by uncontrolled temperature gradients in a cooling assay plate (edge effects), remove cell-based multi-well plate cultures from the incubator and allow for equilibration to room temperature prior to adding the reagent.
2. Intensely colored compounds that absorb light in the emission spectrum of luminescence may reduce the measured light.
3. The possibility that compounds may interfere with the enzymatic activity of luciferase must be considered. Although statistically rare, luciferase inhibition may lead to false negative data.
4. The synchronous dual enzyme mechanism of the assay requires certain kinetic constraints for rigorous determination of potency (IC_{50}).

References

1. Tewari, M., Quan, L.T., O'Rourke, K., Desnoyers, S., Zeng, Z., Beidler, D.R., Poirier, G.G., Salveson, G.S., and Dixit, V.M. (1995) Yama/CPP32, a mammalian homolog of Ced-3, is a CrmA-inhibitable protease that cleaves the death substrate poly(ADP-ribose) polymerase. *Cell* **81**, 801–809.
2. Casciola-Rosen, L.A., Miller, D.K., Anhalt, G.J., and Rosen, A. (1994) Specific cleavage of the 70-kD protein component of the U1 small nuclear ribonucleoprotein is a characteristic biochemical feature of apoptotic cell death. *J. Biol. Chem.* **269**, 30757–30760.
3. Casciola-Rosen, L., Nicholson, D.W., Chong, T., Rowan, K.R. Thornberry, N.A., Miller, D.K., and Rosen, A. (1996) Apopain/CPP32 cleaves proteins that are essential for cellular repair: a fundamental principle of apoptotic death. *J. Exp. Med.* **183**, 1957–1964.
4. Martin, S.J., and Green, D.R. (1995) Protease activation during apoptosis: death by a thousand cuts? *Cell* **82**, 349–352.
5. Chinnaiyan, A.M., and Dixit, V.M. (1996) The cell death machine. *Curr. Biol.* **6**, 555–562.
6. Nicholson, D.W., and Thornberry, N.A. (1997) Caspases: killer proteases. *Trends Biochem. Sci.* **22**, 299–306.
7. Alnemri, E.S., Livingston, D.J., Nicholson, D.W., Salveson, G., Thornberry, N.A., Wong, W.W., and Yaun, J.Y. (1996) Human ICE/CED-3 protease nomenclature. *Cell* **87**, 171–173.
8. Thornberry, N.A., Rano, T.A., Peterson, E.P., Rasper, D.M., Timkey, T., Garcia-Calvo, M., Houtzager, V.M., Nordstrom, P.A., Roy, S., Vaillancourt, J.P., Chapman, K.T., and Nicholson, D.W. (1997) A combinatorial approach defines specificities of members of the caspase family and granzyme B. *J. Biol. Chem* **272**, 17907–17911.

9. Kuida, K., Lippke, J.A., Ku, F., Harding, M.W., Livingston, D.J., Su, M.S., and Flavell, R.A. (1995) Altered cytokine export and apoptosis in mice deficient in interleukin-1 beta converting enzyme. *Science* **267**, 2000–2003.

10. Li, P., Allen, H., Banerjee, S., Franklin, S., Herzog, L., Johnston, C., McDowell, J., Paskind, M., Rodman, L., Salfeld, J., Towne, E., Tracey, D., Wardell, S., Wei, F.-Y., Wong, W., Kamen, R., and Seshadri, T. (1995) Mice deficient in IL-1-converting enzyme are defective in production of mature IL-1 and resistant to endotoxic shock. *Cell* **80**, 401–411.

11. Stennicke, H.R., and Salveson, G.S. (1997) Biochemical characteristics of caspases-3, -6, -7, and -8. *J. Biol. Chem.* **272**, 25719–25723.

12. Garcia-Calvo, M., Peterson, E., Rasper, D., Vaillancourt, J.P., Zamboni, R., Nicholson, D.W., and Thornberry, N.A. (1999) Purification and catalytic properties of caspase family members. *Cell Death. Differ.* **6**, 362–369.

13. Musio, M., Stockwell, B.R., Stennicke, H.R., Salvesen, G.S., and Dixit, V.M. (1998) An induced proximity model for caspase-8 activation. *J. Biol. Chem.* **273**, 2926–2930.

14. Stennicke, H.R., Deveraux, Q.L., Humke, E.W., Reed, J.C., Dixit, V.M., and Salvesen, G.S. (1999) Caspase-9 can be activated without proteolytic processing. *J. Biol. Chem.* **274**, 8359–8362.

15. Siegel, R.M., Frederiksen, J.K., Zacharias, D.A., Chan, F.K., Johnson, M., Lynch, D., Schweizer, A., Briand, C., and Grütterm, M.G. (2003) Crystal structure of caspase-2, apical initiator of the intrinsic apoptotic pathway. *J. Biol. Chem.* **278**, 42441–42447.

16. Chan, F.K., Chun, H.J., Zheng, L., Siegel, R.M., Bui, K.L., and Lenardo, M.J. (2000) A domain in TNF receptors that mediates ligand-independent receptor assembly and signaling. *Science* **288**, 2351–2354.

17. Nicholson, D.W., and Thornberry, N.A. (2003) Life and death decisions. *Science* **299**, 214–215.

18. Tsien, R.Y., and Lenardo, M.J. (2000) Fas preassociation required for apoptosis signaling and dominant inhibition by pathogenic mutations. *Science* **288**, 2354–2357.

19. Lin, C.-F., Chen, C.-L., Chang, W.-T., Jan, M.-S., Hsu, L.-J., Wu, R.-H., Tang, M.-J., Chang, W.-C., and Lin, Y.-S. (2004) Sequential caspase-2 and caspase-8 activation upstream of mitochondria during ceramide and etoposide-induced apoptosis. *J. Biol. Chem.* **279**, 40755–40761.

20. Wagner, K.W., Engels, I.H., and Deveraux, Q.L. (2004) Caspase-2 can function upstream of bid cleavage in the TRAIL apoptosis pathway. *J. Biol. Chem.* **279**, 35047–35052.

21. Slee, E.A., Harte, M.T., Kluck, R.M., Wolf, B.B., Casiano, C.A., Newmeyer, D.D., Wang, H.-G., Reed, J.C., Nicholson, D.W., Alnemri, E.S., Green, D.R., and Martin, S.J. (1999) Ordering the cytochrome c-initiated caspase cascade:

hierarchial activation of caspases-2, -3, -6, -7, -8, and 10 in a caspase-9-dependent manner. *J. Cell Biol.* **137**, 469–479.

22. Liu, J., Bhalgat, M., Zhang, C., Diwu, Z., Hoyland, B., and Klaubert, D.H. (1999) Fluorescent molecular probes V: a sensitive caspase-3 substrate for fluorometric assays. *Bioorg. Med. Chem. Lett.* **9**, 3231–3236.

23. O'Brien, M.A. Daily, W.J., Hesselberth, E.P., Moravec, R.A., Scurria, M., Klaubert, D.H., Bulleit, R.F., and Wood, K.V. (2005) Homogeneous, bioluminescent protease assays: caspase-3 as a model. *J. Biomol. Screen.* **10**, 137–148.

24. Riss, T.L., and Moravec, R.A. (2004) Use of multiple assay endpoints to investigate the effects of incubation time, dose of toxin, and plating density in cell-based cytotoxicity assays. *Assay Drug Dev Technol* **2**, 51–62.

12

Multiplex Caspase Activity and Cytotoxicity Assays

Andrew L. Niles, Richard A. Moravec, and Terry L. Riss

Summary

Multiplexed assay chemistries provide for multiple measurements of cellular parameters within a single assay well. This experimental practice not only is more cost efficient but provides more informational content about a compound or treatment. For instance, multiplexed caspase activity assays can help establish the kinetics and magnitude of initiator and effector caspase induction by candidate compounds or treatments. The ability to combine the activity profiles within the same sample provides a level of normalization not possible with parallel assays. Furthermore, multiplexing caspase activity assays with viability and/or cytotoxicity assays can support conclusions regarding cytotoxic mechanism and provide normalization that may help correct for differences in cell number.

Key Words: Multiplex; fluorescence; luminescence; cell based; cytotoxicity; caspase.

1. Introduction

The term "multiplex" is used extensively in biology to describe techniques or assays that capture more than one set of data from the same sample by measuring different parameters. The motivation for combining assays within the same well is twofold: reduction in cost of reagents, consumables, and operator time versus parallel assays and the intrinsic power of intra-well response normalization. Regardless of which specific experimental application is being described, all multiplex assays have an obvious requirement for the combined assay chemistries to be compatible, distinct, and measurable. Assay signal separation is achieved by various means including using fluorophores with divergent excitation and emission spectra (*1*), using chemiluminescence

From: *Methods in Molecular Biology, vol. 414: Apoptosis and Cancer*
Edited by: G. Mor and A. B. Alvero © Humana Press Inc., Totowa, NJ

and bioluminescence *(2)*, using fluorescence and bioluminescence *(3)*, or using bioluminescence in a sequential manner with the aid of a quenching agent *(4)*. For the purposes of this chapter, we will describe only those chemistries germane to cytotoxicity, viability, and caspase activation assays.

1.1. Multiplexed Caspase Activity Assays

Caspase activation in cell culture is an early and definitive hallmark of apoptosis. The ability to further characterize the caspase induction response is useful for profiling a compound or treatment *(5)*. Multiplexed caspase activity assays have utility in delineating the functional and temporal aspects of caspase activation when used in time-course experiments. For instance, the initiator caspases 8 and 9 are activated by specific extrinsic or intrinsic stimuli, respectively. In turn, these caspases activate the effector caspases (3, 6, and 7) that mediate cellular destruction (see 11 chapter for caspase biology and apoptosis process review). Therefore, carefully configured homogeneous multiplexed caspase activity assays can offer insight into the response magnitude, kinetics, and primary pathway relative to model inducers.

Combining caspase detection chemistries is simple if standard reaction conditions for both enzymes are met. The reagent formulations in most commercial kits designed to measure individual caspase activities contain optimized concentrations of the buffering agent, salts, thiols, and cofactors, all at a specific pH. Practically speaking however, optimized buffer formulation differences are typically minimal, and the caspase family as a whole are generally accommodating of these subtle differences *(6)*. Therefore, an optimal buffer system for one enzyme may be adequate for assaying a range of different caspase activities.

The primary basis for selectivity in caspase activity assays is imparted by the tetrapeptide sequence of the substrate. The secondary basis for selectivity is the kinetic constants of the substrate with respect to the target and off-target caspases. When attempting to measure more than one (multiple) caspase activities from a mixture, it is essential that the substrates are added to the sample at appropriate concentrations. The most useful combinations of substrate concentrations often reflect the individual K_m of targeted enzymes.

1.2. Cytotoxicity and/or Viability Assays Multiplexed with Caspase Activity Assays

Experimental manipulation of cells in culture leads to three general outcomes after a defined exposure: no effect (with regard to viability or cytotoxicity), proliferation, or cytotoxicity. Viability and/or cytotoxicity assays are therefore useful in defining either the general tolerability of a test compound or the demonstrating cytotoxic potential.

It is often useful to couple the primary response data (e.g., caspase activity) with global changes in cellular health when attempting to identify compounds with apoptosis-inducing activity. For instance, caspase activation with a commensurate reduction in viability (or increase in cytotoxicity) relative to the control treatment would strongly support the conclusion of programmed cell death by apoptosis. However, cytotoxicity and/or reduction in cellular viability in the absence of caspase activation may indicate a primary necrotic event. The ability to distinguish between these forms of cytotoxicity in a plate-based format by a homogeneous multiplexed assay is often relevant from a therapeutic perspective, because apoptosis can often be modulated, whereas the necrotic process is less understood and typically detrimental *(7)*.

The quality of caspase activity or inhibition data often can be improved by determining the relative number of live or dead cells remaining at an endpoint after treatment *(8)*. For instance, a single parameter assay measuring only caspase-3/7 activation may skew or confound the data set if cell clumping or other experimental error influenced the initial number of cells in the sample well. The risk is that statistically significant increases or decreases in caspase activity may be due to cell number alone, not a specific biological effect. Therefore, these multiplexed methods can substantially improve and control erroneous conclusions and variation by response normalization.

Procedurally, multiplexed viability and/or cytotoxicity assays are easily configured by the sequential addition of viability and/or cytotoxicity reagents to the assay well prior to homogeneous caspase reagent addition *(9,10)*. This sequence-dependent addition is necessary because homogeneous caspase activity assay buffers contain lytic components and reducing thiols in their buffer systems, which would effect cellular viability and interfere with viability or cytotoxicity measurements.

2. Materials

2.1. Equipment

1. 15-ml conical tubes (Corning Lowell, MA, USA).
2. 1.5-ml Eppendorf tubes.
3. 75 cm^2 culture flasks (Corning).
4. Reagent reservoirs (Corning).
5. Sterile pipette tips.
6. Single and multi-channel micro-pipettors.
7. Hemocytometer and coverslip.

8. Bright-field microscope.
9. Table-top centrifuge.
10. Opaque multi-well plate (Corning).
11. Multi-well fluorometer equipped with the following filter pairs:

   a. $400_{ex} \cdot 505_{em}$
   b. $485_{ex} \cdot 527_{em}$
   c. $560_{ex} \cdot 590_{em}$.

12. Multi-well luminometer.
13. CO2 incubator.

2.2. Reagents

1. Cell-culture medium (RPMI, MEM, DMEM, etc.).
2. Animal serum (fetal bovine, horse, etc.).
3. HL-60 cell line (ATCC Manassas, VA, USA).
4. HepG2 cell line (ATCC).
5. Jurkat cell line (ATCC).
6. Trypan blue solution (Sigma, St. Louis, MO, USA).
7. Tamoxifen (Sigma).
8. Dimethyl sulfoxide (DMSO) (Sigma).
9. Recombinant TRAIL (Chemicon Temecula, CA, USA).
10. Apo-ONE® Homogeneous Caspase-3/7 Assay (Promega Madison, WI, USA).
11. Caspase-Glo®-3/7, 8, and 9 Activity Assays (Promega).
12. CellTiter-Blue® Cell Viability Assay (Promega).
13. MultiTox-Fluor Multiplex Cytotoxicity Assay (Promega).

3. Methods

The methods described below outline experimental examples of (i) a caspase induction time course in cell culture, (ii) simultaneous multiplexed caspase measurement by a luminescent and fluorescent activity assay, (iii) dose-dependent apoptosis, (iv) measurement of apoptotic responses by a multiplexed fluorescent viability and caspase-3/7 assay, or (v) a multiplexed fluorescent cytotoxicity and luminogenic caspase activity assay.

3.1. A Caspase Induction Time Course

Caspase activation can be initiated by a multitude of stimuli *(11)*. However, the resulting caspase responses can be functionally separated into two primary pathways based on apoptotic mechanism. For instance, cell death signals that stimulate the "intrinsic" mitochondrial pathway recruit and activate caspase-9 by forming a multi-unit complex known as the apoptosome. Activated caspase-9 can then act on pro-caspase-3 to affect caspase-3 activation and, ultimately, the executioner function. The "extrinsic" pathway is characterized by utilizing cell surface receptors to engage cell death ligands that transduce signals for caspase-8 to associate with adaptor molecules. Activated caspase-8 can then activate pro-caspase-3 in a manner like caspase-9.

The following protocol induces an extrinsic caspase activation profile by using tumor necrosis factor-related apoptosis-inducing ligand (TRAIL). TRAIL receptor expression is variable among cells of different lineages with primary lines being particularly insensitive to induction *(12)*. The protocol can be amended to use staurosporine to promote the intrinsic activity profile.

To avoid culture contamination by microorganisms, conduct cell-based assay experiments in a laminar flow hood or clean-room environment using aseptic technique with sterile reagents and consumables.

1. Harvest cells from culture. Remove a representative volume (1000 μl or less) for analysis. Gently pellet the remaining pool by centrifugation at $200 \times g$ for 8–10 min at room temperature.
2. Examine a dilution of the cell sample by trypan blue exclusion to determine the population viability and cell count. If viability is greater than 90%, add complete medium (with serum and nutrient/cofactor adjuncts) so that the cells are at a density of 500,000 viable cells/ml.
3. Plate the cells in 50 μl volumes (25,000 cells/well) to a sterile, opaque-walled microtiter plate. Add 100 μl volumes of cell-culture medium to replicate wells to serve as cell-free control. If the cells are attachment dependent, allow for a 2- to 16-h equilibration period before proceeding.
4. Prepare caspase induction control agent. For example, dilute rTRAIL to 200 ng/ml in complete medium. Add the control-inducing agent in replicate 50 μl volumes to replicate cell wells every hour for a 10-h time course.
5. Dilute unknown or test compound to 20 μM in complete medium and add them in 50 μl volumes to cell wells over the same time course.
6. Add 50 μl complete cell-culture medium (vehicle) or culture medium with dilution-matched solvent (e.g., 0.1% DMSO, when appropriate) to wells designated for non-induction, negative control.
7. Mix the plate by orbital shaking after each addition to ensure homogeneity and compound dispersion.
8. Incubate the plate at 37°C in a humidified CO_2 environment over the experimental time course.
9. Proceed to multiplex caspase activity assay (*see* **Subheading 3.2.**).

3.2. Multiplex Caspase Assay

The biochemical basis for homogeneous, plate-based measurement of active caspases has been described in greater detail in a previous chapter 11 in this

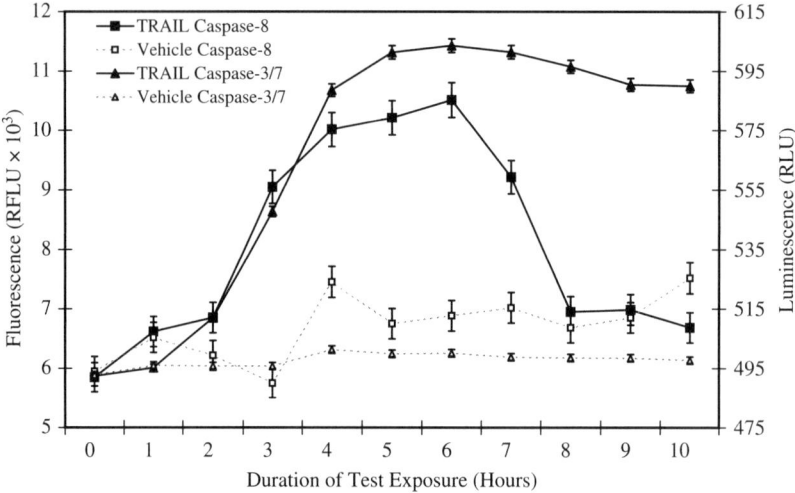

Fig. 1. Multiplex luminescent caspase-8 and fluorescent caspase-3/7 assays of rTRAIL-treated HL-60 cells. HL-60 cells were seeded at 25,000 cells/well in 50 μl RPMI 1640 + 10% FBS. 50 μl rTRAIL (Chemicon, 100 ng/ml final) in RPMI 1640 or vehicle control (RPMI 1640) was added to replicate wells every hour over a 10-h time course. Caspase-Glo®-8 Reagent (Promega) was prepared by adding the provided assay buffer to the substrate cake. A fluorescent caspase-3/7 substrate (bis-Z-DEVD-R110) from the Apo-ONE® kit (Promega) was mixed into the Caspase-Glo®-8 reagent at a final concentration of 50 μM. 100 μl combined reagent was added to the cells and incubated 60 min prior to measuring luminescence and fluorescence using the BMG FluoStar™. A.L. Niles, unpublished data.

book. This protocol merges chemistries supporting luminescent detection of a caspase activity with a fluorescent caspase detection substrate. The resulting signals for the respective caspase activities can be spectrally separated by using a fluorometer and luminometer.

The following is an abbreviated protocol for combining a luminescent caspase-8 or caspase-9 assay with a fluorescent caspase-3/7 substrate. For details of the assay formats and potential usages, consult the Promega publication *(24)* that describes the complete series of multiplexed applications.

1. Equilibrate the frozen Caspase-Glo®-8 (or Caspase-Glo®-9) buffer and lyophilize Caspase-Glo®-8 (or Caspase-Glo®-9) substrate to room temperature prior to use.
2. Transfer the Caspase-Glo®-8 (or Caspase-Glo®-9) buffer into the lyophilized Caspase-Glo®-8 (or Caspase-Glo®-9) substrate to create the reagent. Mix by swirling or by inverting the contents until the substrate is completely dissolved.

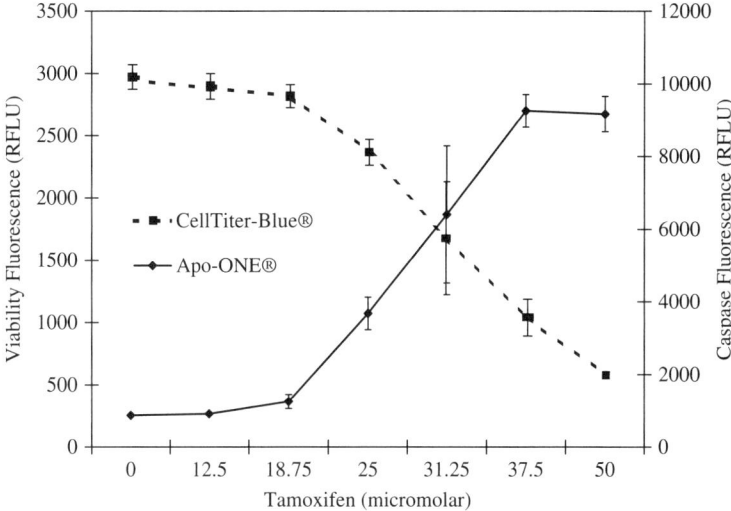

Fig. 2. Multiplex viability and caspase-3/7 assays of tamoxifen-treated HepG2 cells. HepG2 cells were seeded at 10,000 cells/well in 50 μl MEM + 10% FBS and allowed to attach. Serial dilutions of tamoxifen (Sigma) were made in MEM + 10% FBS and added in 50 μl. A DMSO-matched vehicle served as control. The compound dilution series was incubated for 3.5 h. 5 μl CellTiter-Blue® Reagent (Promega) was added to the wells and incubated an additional hour at 37°C. The plate was then removed and an equal volume of Apo-ONE® Reagent added to each well. Fluorescence was measured after 30 min. R.A. Moravec, unpublished data.

3. Remove the micro-well plate containing samples and controls from the incubator and equilibrate to room temperature.
4. Thaw the 100× substrate from the Apo-ONE® kit and admix it into the Caspase-Glo® reagent at a dilution ratio of 1:100.
5. Add the reagent mixture at a ratio of 1:1 (e.g., 100 μl reagent to 100 μl sample or control), mix by orbital shaking, and incubate for 1 h.
6. Measure luminescence using a luminometer, then measure the fluorescent product accumulation using a fluorometer with excitation in the range of 485 ± 20 nm and an emission filter at 530 ± 20 nm.

3.3. Dose-Dependent Apoptosis in a Cell-Culture Model

1. Harvest and prepare HepG2 (or Jurkat) cells (*see* **Subheading 3.1.**, steps 1 and 2) in complete medium with the exception of re-suspending the cells at a final density of 200,000 viable cells/ml.

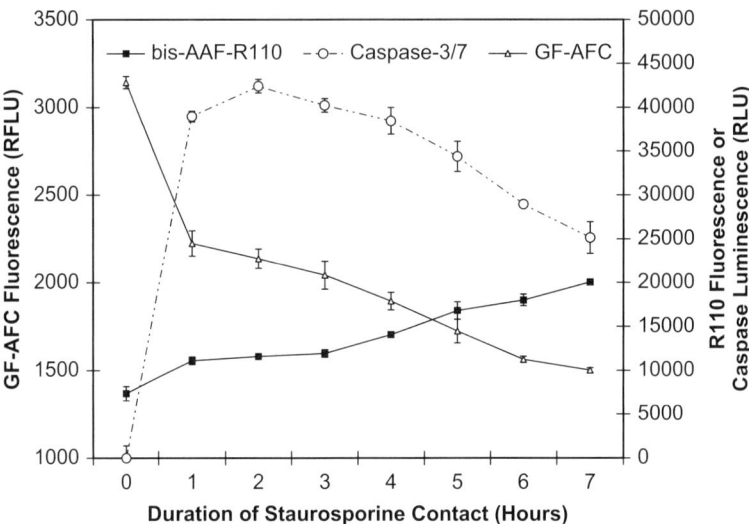

Fig. 3. Multiplexed viability and cytotoxicity with caspase-3/7 activity assay. Jurkat were seeded at 10,000 cells/well in 50 µl volumes and allowed to equilibrate for 1 h. Staurosporine was diluted to 10 µM in RPMI 1640 + 10% FBS and added to replicate wells every hour over a 7-h time course. RPMI 1640 + 10% FBS with staurosporine-matched DMSO concentrations were added at the same time to serve as vehicle controls. MultiTox-Fluor Reagent was added to the wells in a 10 µl volume and viable and cytotoxic populations measured by fluorescence after 30 min of incubation at 37°C. Caspase-Glo®-3/7 Reagent was added and luminescence measured after 20 min. The raw fluorescence data for the viability and cytotoxicity assay is plotted versus the background subtracted (vehicle + cells) luminescence values. A.L. Niles, unpublished data.

2. Plate the cells in 50 µl (10,000 cells/well) to a sterile, opaque-walled micro-well plate. Add 100 µl cell-culture medium to replicate wells to serve as cell-free control. If the cells are attachment dependent, allow for a 2- to 16-h equilibration period before proceeding.
3. Prepare a caspase induction control agent. For example, dilute tamoxifen to 100 µM in complete medium. Twofold serially dilute the compound in complete medium in a sterile micro-well plate using a multi-channel pipettor. Add the control-inducing agent dilutions in replicate 50 µl volumes to replicate cell wells in the test plate.
4. Dilute additional test compound to 20 µM in complete medium and then dilute and add them in 50 µl volumes to cell wells as described for the control agent.
5. Add 50 µl complete cell-culture medium (vehicle) or culture medium with dilution-matched solvent (e.g., 0.1, 0.05% DMSO, etc., when appropriate) to wells designated for non-induction, negative control.

6. Mix the plate by orbital shaking after each addition to ensure homogeneity and compound dispersion.
7. Incubate the plate at 37°C in a humidified CO2 environment for a period between 3.5 and 24 h.
8. Proceed to protocols either 3.4 or 3.5.

3.4. Multiplex Viability and Caspase-3/7 Assay

Some cell viability assay chemistries are designed to measure biochemical events that occur in viable but not compromised or dead cells. One such measure of viability is metabolic capacity. In the following assay, a resazurin dye is introduced into the cell sample in a physiological formulation. Normal viable cells reduce the dye in a proportional manner, and the resulting product (resorufin) can be measured with a fluorometer. Therefore, decreases in fluorescence compared with control indicate cytotoxicity. After the viability measurement has been recorded, the caspase activity assay reagent can be added to the sample well to report caspase activities.

The following is an abbreviated protocol for performing a multiplexed viability and homogeneous caspase assay to measure caspase-3/7 activity from cells in culture. For complete details of both assay formats and potential usages, consult Promega Technical Bulletin #295 that describes the Apo-ONE® Homogeneous Caspase-3/7 Assay and Promega Technical Bulletin #317 that describes the CellTiter-Blue® Reagent.

1. Thaw the CellTiter-Blue® Reagent and equilibrate to ambient temperature.
2. Add CellTiter-Blue® Reagent to each sample well during the last 2 h of experimental treatment (5 μl reagent per 100 μl sample). Gently mix by orbital shaking to ensure that the reagent is equally dispersed. Return the plate to 37°C and incubate for 1–3 h.
3. Thaw Apo-ONE® Assay buffer and 100× substrate and allow them to equilibrate to room temperature. Mix by manual inversion or by vortexing.
4. Dilute Apo-ONE® Substrate 1:100 with the buffer to obtain the desired volume of Apo-ONE® Homogeneous Caspase-3/7 Reagent.
5. At the end of the CellTiter-Blue® Reagent incubation period, remove the plate from incubator and measure fluorescence at an excitation of 560 nm and emission of 590 nm.
6. Add the Apo-ONE® Reagent to wells of the plate at a ratio of 1:1 (e.g., 105 μl Apo-ONE® Reagent to 100 + 5 μl CellTiter-Blue® Reagent). Mix by orbital shaking to ensure homogeneity and incubate at room temperature for at least an hour prior to recording fluorescence at an excitation of 485 ± 20 nm and emission of 530 ± 20 nm.

3.5. Multiplex Caspase Activity and Cytotoxicity Assay

Measurement of cell membrane integrity is a common method to establish the degree of cytotoxicity or viability within a population in culture. The MultiTox-Fluor assay used in this protocol simultaneously measures two protease activities: one is a marker of cell viability and the other is a marker of cytotoxicity. The live cell protease activity is restricted to intact viable cells and is measured using a fluorogenic, cell permeable, peptide substrate. The substrate enters intact cells where it is cleaved to generate a fluorescent signal proportional to the number of living cells. This live cell protease activity marker becomes inactive upon loss of membrane integrity and leakage into the surrounding culture medium. A second, cell impermeant, fluorogenic peptide substrate is used to measure dead cell protease activity that has been released from cells that have lost membrane integrity. By using this assay, the live and dead cell populations are directly defined in a non-lytic, homogeneous format. Furthermore, this inversely complimentary viability profile provides a ratiometric response that can improve assay precision and reduce erroneous interpretation of the results.

The following is an abbreviated protocol for performing a multiplexed cytotoxicity and homogeneous caspase assay to measure caspase-3/7 activity from cells in culture. For complete details of the both assay formats and potential usages, consult Promega Technical Bulletin #348 that describes the MultiTox-Fluor Multiplex Cytotoxicity Assay and Promega Technical Bulletin #323 that describes the Caspase-Glo®-3/7 Assay.

1. Thaw the MultiTox-Fluor substrates and assay buffer components.
2. Vortex the components to ensure homogeneity.
3. Add 980 μl assay buffer to a 1.5-ml Eppendorf tube. Create a 20× MultiTox-Fluor Reagent by adding 10 μl GF-AFC and 10 μl bis-AAF-R110 to the assay buffer tube. Vortex vigorously.
4. With a multi-channel pipettor, add 10 μl 20× MultiTox-Fluor Reagent to all wells of the plate. Mix by orbital shaking, then return the plate to the 37°C incubator for at least 30 min, but not longer than 3 h.
5. Equilibrate the frozen Caspase-Glo®-3/7 buffer and lyophilize Caspase-Glo®-3/7 substrate to room temperature prior to use.
6. Measure fluorescence associated with viable and cytotoxic populations using the following optimal wavelengths: $400_{ex}\cdot505_{em}$ (live cells) and $485_{ex}\cdot527_{em}$ (dead cells).
7. Transfer the Caspase-Glo®-3/7 buffer into the lyophilized Caspase-Glo®-3/7 substrate to create the reagent. Mix by swirling or by inverting the contents until the substrate is completely dissolved.
8. After 20–30 min of incubation at room temperature, measure the luminescent signal associated with caspase-3/7 activity.

4. Notes

1. The activity profiles for different caspases change throughout the apoptotic program. Similarly, the kinetics and potency of any treatment may vary with treatment or compound contact time *(13)*. Multiple measurements at different time points may be necessary to define the peak response for initiator or effector caspases. Single-point multiplexed analyses may underestimate or miss these responses.
2. Caspase activity feedback loops may confound the interpretation of initiation profiles *(14)*.
3. When multiplexing a viability or cytotoxicity assay with a caspase activity measure, always follow a sequential order of assay chemistry addition. This is required because homogeneous caspase assay formulations contain agents that lyse cells. Therefore, addition of caspase reagents prior to viability or cytotoxicity reagents would automatically kill all cells in the culture.
4. Normalization of caspase activity responses to cell number by viability or cytotoxicity assays is particularly useful in multi-well formats. However, care should be taken in such normalization because caspases have activity half-lives and caspase activity might be measurable prior to changes in cellular viability.
5. The CellTiter-Blue® chemistry utilizes the reduction of resazurin dye into the fluorescent product resorufin. The intense color of the dye can lead to quenching of the signal from either fluorescent or luminescent caspase activity assays. Although thiols present in caspase activity buffers typically reduce remaining resazurin to resorufin with short incubations, it is necessary to verify the uniformity in color prior to measuring fluorescence or luminescence.

References

1. Grant, S., Sklar, J., and Cummings, R. (2002) Development of novel assays for proteolytic enzymes using rhodamine-based substrates. *J. Biomol. Screen.* **7**, 531–540.
2. Bronstein, I., Fortin, J., Stanley, P. E., Stewart, G. S. A. B., and Kricka, L. J. (1994) Chemiluminescent and bioluminescent reporter gene assays. *Anal. Biochem.* **219**, 169–181.
3. Wesierska-Gadek, J., Gueorguieva, M., Ranftler, C., and Zerza-Schnitzhofer, G. (2005) A new multiplex assay allowing simultaneous detection of the inhibition of ccll proliferation and induction of cell death. *J. Cell Biochem.* **96**, 1–7.
4. Nieuwenhuijsen, B., Huang, Y., Wang, Y., Ramerez, F., Kalgaonkar, G., and Young, K. (2003) A dual luciferase multiplex high-throughput screening platform for protein-protein interactions. *J. Biomol. Screen.* **8**, 676–684.
5. Niles, A., Moravec, R., and Riss, T. (2004) Characterizing responses to treatments using homogeneous caspase activity and cell viability assays. *Cell Notes* **9**, 11–14.
6. Garcia-Calvo, M., Peterson, E., Rasper, D., Vaillancourt, J. P., Zamboni, R., Nicholson, D. W., and Thornberry, N. A. (1999) Purification and catalytic properties of caspase family members. *Cell Death Differ.* **6**, 362–369.

7. Leist, M., and Jaattela, M. (2001) Four deaths and a funeral: from caspases to alternative mechanisms. *Nat. Rev. Mol. Cell. Biol.* **2**, 589–598.

8. Niles, A., Worzela, T., Scurria, M., Daily, W., Bernad, L., Guthmiller, P., McNamara, B., Rashka, K., Lange, D. and Riss, T. (2006) Multiplexed viability, cytotoxicity and apoptosis assays for cell-based screening. *Cell Notes* **16**, 12–15.

9. Niles, A., Moravec, R., Scurria, M., Daily, W., Bernad, L., McNamara, B., Moraes, A., Rashka, K., Lange, D., and Riss, T. (2006) MultiTox-fluor multiplex cytotoxicity assay technology. *Cell Notes* **15**, 11–15.

10. Farfan, A., Yeager, T., Moravec, R., and Niles, A. (2004) Multiplexing homogeneous cell-based assays. *Cell Notes* **10**, 15–17.

11. Nicholson, D., and Thornberry, N. (2003) Life and death decisions. *Science* **299**, 214–215.

12. Ashkenazi, A., Pai, R., Fong, S., Leung, S., Lawrence, D., Marsters, S., Blackie, C., Chang, L., McMurtrey, A., Hebert, A., DeForge, L., Koumenis, I., Lewis, D., Harris, L., Bussiere, J., Koeppen, H., Shahrokh, Z., and Schwall, R. (1999) Safety and antitumor activity of recombinant soluble Apo2 ligand. *J. Clin. Invest.* **104**, 155–162.

13. Riss, T., and Moravec, R. (2004) Use of multiple assay endpoints to investigate the effects of incubation time, dose of toxin, and plating density in cell-based assays. *Assay Drug Dev. Technol.* **2**, 51–62.

14. Roy, S., and Nicholson, D. (2000) Cross-talk in cell death signalling. *J. Exp. Med.* **192**, 21–25.

13

Homogeneous, Bioluminescent Proteasome Assays

Martha A. O'Brien, Richard A. Moravec, Terry L. Riss, and Robert F. Bulleit

Summary

Protein degradation is mediated predominantly through the ubiquitin–proteasome pathway. The importance of the proteasome in regulating degradation of proteins involved in cell-cycle control, apoptosis, and angiogenesis led to the recognition of the proteasome as a therapeutic target for cancer (*1–4*). The proteasome is also essential for degrading misfolded and aberrant proteins, and impaired proteasome function has been implicated in diseases such as Parkinson's and Alzheimer's (*5*). The importance of the proteasome for general cell homeostasis has been established, and the 2004 Nobel Prize for Chemistry honored the researchers that discovered the ubiquitin–proteasome pathway. Robust, sensitive assays are essential for monitoring proteasome activity and for developing inhibitors of the proteasome. Peptide-conjugated fluorophores are widely used as substrates for monitoring proteasome activity, but fluorogenic substrates can exhibit significant background and can be problematic for screening because of cellular autofluorescence or fluorescent library compounds. To address these issues, we developed a homogeneous, bioluminescent method that combines peptide-conjugated aminoluciferin substrates and a stabilized luciferase. We have developed homogeneous, bioluminescent assays for all three proteasome activities, the chymotrypsin-like, trypsin-like, and caspase-like, using purified proteasome. We have also applied this technology to a cellular assay using the substrate for the chymotrypsin-like activity in combination with a selective membrane permeabilization step (patent pending). The proteasome assays are designed in a simple "add and read" format and have been tested in 96- and 384-well plates. The bioluminescent, coupled-enzyme format enables sensitive and rapid protease assays ideal for inhibitor screening.

Key Words: 20S proteasome; bioluminescence; luciferase; aminoluciferin; bioluminescent protease assay; 20S proteasome assay.

From: *Methods in Molecular Biology, vol. 414: Apoptosis and Cancer*
Edited by: G. Mor and A. B. Alvero © Humana Press Inc., Totowa, NJ

1. Introduction

1.1. The Ubiquitin–Proteasome Pathway

In eukaryotic cells, the turnover of intracellular proteins is mediated mainly by the ubiquitin–proteasome pathway, a non-lysosomal proteolytic pathway. The 26S proteasome is a 2.5-MDa multiprotein complex found in both the nucleus and the cytosol of all eukaryotic cells and is comprised of a single 20S core particle and 19S regulatory particles at one or both ends *(6,7)*. Three major proteolytic activities are contained within the 20S core. Together these three activities are responsible for much of the protein degradation required to maintain cellular homeostasis including degradation of critical cell-cycle proteins, tumor suppressors, transcription factors, inhibitory proteins, and damaged cellular proteins *(8,9)*. Proteins destined to be degraded by the proteasome are first selectively targeted by the addition of a series of covalently attached ubiquitin molecules. The 26S proteasome degrades poly-ubiquitinated proteins in an ATP-dependent manner. The 19S regulatory unit binds and removes the ubiquitin chains from tagged proteins, and ATPases within the regulatory complex appear to unfold protein substrates and translocate the unfolded polypeptides into the 20S core *(8,9)*. There the polypeptides are degraded to yield peptides ranging from 3 to 25 amino acids in length *(10)*. The 20S catalytic core and the 19S regulatory complex are highly conserved from yeast to mammals *(11)*.

The catalytic core of the complex, the 20S proteasome, is a barrel-shaped assembly of 28 protein subunits that possesses three different proteolytic activities designated as chymotrypsin-like, trypsin-like, and caspase-like (also termed post-glutamyl peptide hydrolase) *(11,12)*. The catalytic sites are located on the inner surface of the central β-rings of the cylindrical particle, and access to them is controlled by narrow, gated channels in the outer α-rings of the complex. The association of the 20S particle with a 19S regulatory complex at one or both ends of the barrel forms the 26S proteasome and confers an open-channel conformation, resulting in much higher rates of peptide hydrolysis in the 26S proteasome *(13,14)*. Robust, sensitive assays for the catalytic activities of the proteasome will aid in the discovery of new inhibitors.

The proteasome has been validated as a therapeutic target for cancer treatment. Proteasome inhibitors can induce apoptosis, and interestingly, some transformed cells display greater susceptibility to proteasome inhibition than non-malignant cells *(3)*. The enhanced proliferative rate of malignant cells may cause accumulation of damaged proteins at a higher rate that in turn would increase dependency on proteasomal degradation *(15)*. The first-generation proteasome inhibitor, bortezomib (PS-341, Velcade®), is now an FDA-approved drug for the treatment of refractory multiple myeloma, and second-generation inhibitors are currently being developed *(16)*. The clinical testing of bortezomib, as well as other new proteasome inhibitors, for efficacy on an array of cancers is currently in progress *(17,18)*. We describe here the in vitro and cell-based bioluminescent proteasome assays, demonstrate their utility, and compare them with fluorescent assays. These assays enable robust, sensitive and rapid monitoring of proteasome catalytic activities in a simple multiwell format.

1.2. Bioluminescent Proteasome Assay Concept

Peptide-conjugated fluorophores are widely used as substrates for monitoring proteasome activity, but sensitivity of fluorescent assays can be limited for a variety of reasons. Peptide-conjugated fluorophores can have residual fluorescence or spectral overlap with their cleaved fluorescent products, thus increasing background and reducing sensitivity *(19,20)*. Cells can exhibit autofluorescence and compounds from natural product or synthetic chemical libraries frequently exhibit fluorescence that can cause assay interference *(21)*. To provide an alternative to fluorescence, we synthesized luminogenic substrates and developed a homogeneous method for monitoring proteasome activity. We first developed this coupled-enzyme bioluminescent method for monitoring caspase activities *(22)* and have now developed bioluminescent assays for all three proteasome activities. Using standard Fmoc chemistry, we synthesized luminogenic versions of the commonly used fluorogenic coumarin-based substrates, Suc-LLVY-aminoluciferin, Z-LRR-aminoluciferin, and Z-nLPnLD-aminoluciferin, to monitor the chymotrypsin-like, trypsin-like, and caspase-like activities of the proteasome, respectively (*see* **Fig. 1**).

The bioluminescent assays are homogeneous assays, such that the proteasome and luciferase function simultaneously. As a result of this coupled-enzyme format, the proteasome and luciferase rapidly reach a steady-state, where the rate of proteasome cleavage of the substrate is equal to the rate of luciferase utilization of the released aminoluciferin, and stable light output is achieved. Steady-state is typically reached in 10–20 min, and stable light output persists for several hours (*see* **Fig. 2**). Eventually, the light output decreases when the 20S proteasome and the luciferase become inactivated, but the half-life for each of the proteasome assays is greater than 3 hours (*see* **Fig. 2**).

At steady-state, the light output is proportional to the rate of proteasome cleavage and thus the amount of proteasome activity (*see* **Fig. 3**). Another

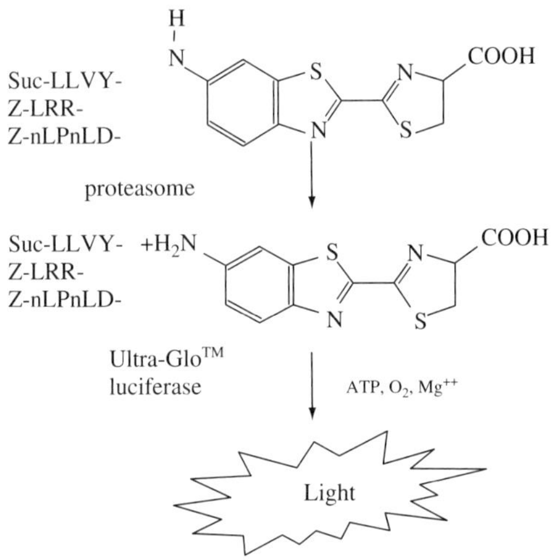

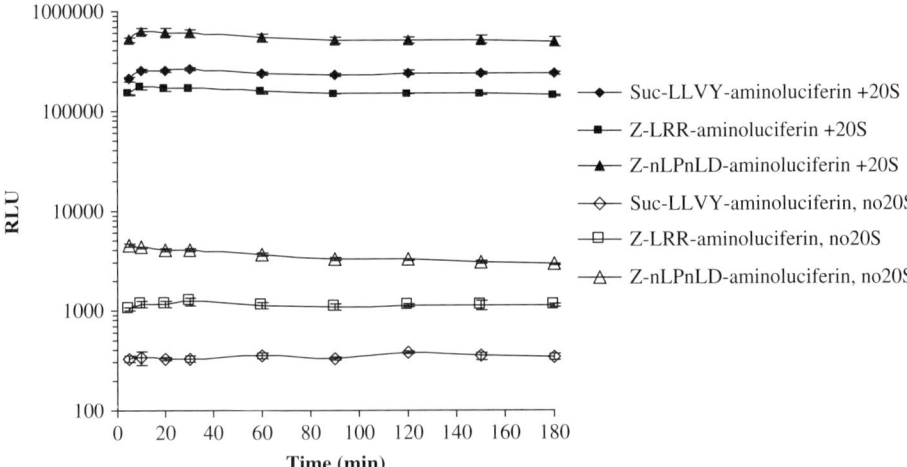

Fig. 1. The luminogenic, aminoluciferin substrates containing the Suc-LLVY, Z-LRR, or Z-nLPnLD sequence recognized by 20S proteasome. Following 20S proteasome cleavage, the substrate for luciferase (aminoluciferin) is released, allowing the luciferase reaction to produce light.

Fig. 2. Signal stability of the proteasome assays. The proteasome assays (Proteasome-Glo™ 3-Substrate System) were tested with human purified 20S proteasome (1 μg/ml) (closed symbols) or without 20S proteasome as a control (open symbols) in 96-well plates in 100 μl total volume. Luminescence was monitored at various times for 3 h on a Glo-Max™ 96-Microplate luminometer. The signals peak rapidly and then are very stable for all three assays as shown on a log scale.

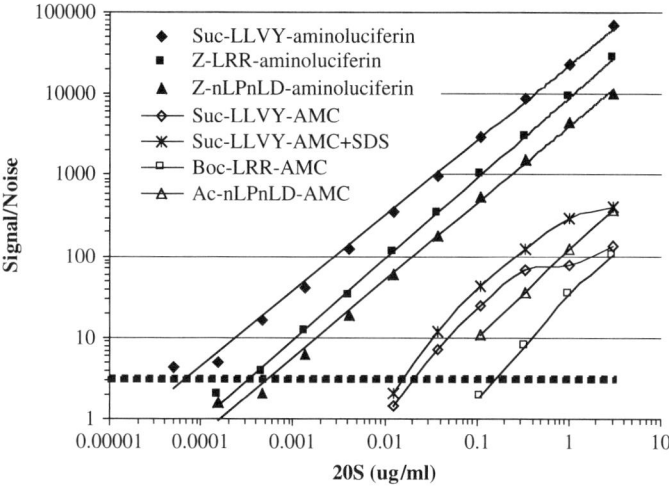

Fig. 3. Luminescent proteasome assays are more sensitive than fluorescent proteasome assays. Human 20S proteasome was serially diluted in 10 mM HEPES (pH 7.6) in 96-well plates. For each catalytic activity, half the plate received the appropriate Proteasome-Glo™ Reagent and half the plate received the comparable fluorogenic substrate, diluted in 100 mM HEPES, pH 7.5, 1 mM EDTA, to the same concentration as the luminescent substrates. The fluorescent assay for chymotrypsin-like proteasome activity was run with and without 0.02% SDS (*see* **Note 1**). Thirty minutes after addition of the Proteasome-Glo™ Reagent, luminescence was recorded as relative light units (RLU) on a Glo-Max™ 96-Microplate luminometer. Fluorescence was measured 30 min after adding the appropriate substrate on a LabSystems Fluoroskan Ascent fluorometer and recorded as relative fluorescence units (RFU). To normalize between RLU and RFU, the results were plotted as signal to noise ratios. Each point represents the average of four wells. The Proteasome-Glo™ Assays were linear over four logs of 20S proteasome concentration for all three assays. The limit of detection is defined as a signal to noise ratio = 3 (dotted line). The luminescent proteasome assays give higher signal to noise ratios and lower limits of detection than the fluorescent assays. SDS, which is frequently used to activate the proteasome when measuring the chymotrypsin-like activity, improved the signal to noise ratio and linearity of the fluorescent assay using Suc-LLVY-AMC, but the sensitivity still did not approach that of the luminescent assay.

feature of the homogeneous, bioluminescent format is that any free aminoluciferin that is a by-product of the peptide-conjugating synthesis is removed prior to exposing the proteasome substrate to the test samples. Consequently, the background is very low, and the linear dynamic range is very large (*see* **Fig. 3**).

The broad dynamic range and stable signal results in increased sensitivity and flexibility for the bioluminescent proteasome assays. A comparison of biolumi-nescent and fluorescent proteasome assays demonstrates that the bioluminescent assays are significantly more sensitive and have a much lower limit of detection for proteasome activity (*see* **Fig. 3**).

1.3. Cellular Bioluminescent Proteasome Assay Concept

Typically, proteasome activity is measured in cells by making cell lysates, using various methods to enrich for proteasome, and then testing for activity using fluorogenic substrates *(23)*. Being able to monitor proteasome directly in cells in multiwell culture dishes would have obvious advantages for high-throughput screening applications. The sensitivity of the bioluminescent, homogeneous format enabled the development of a direct cellular assay for proteasome activity. In addition to sensitivity, specificity is critical when devel-oping a cellular assay. The proteasome catalytic activities are described as

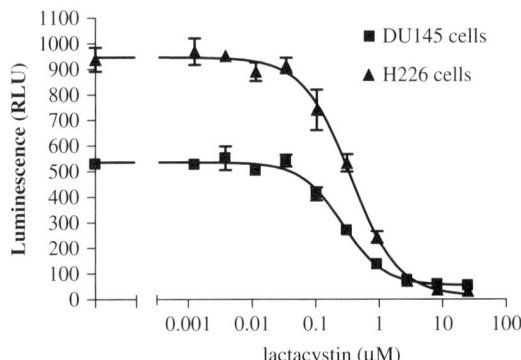

Fig. 4. Lactacystin inhibits the proteasome activity in cultured cells. The Proteasome-Glo™ Cell-Based Assay was used to generate inhibition curves using DU145 and H226 cells. DU145 (human prostate-derived cells grown in MEM containing 10% FBS, 1 mM sodium pyruvate, and 1× non-essential amino acids) and H226 (human lung-derived cells grown in RPMI-1640 containing 10% FBS and 1 mM sodium pyruvate) cells were plated at 5000 cells per well and 2500 cells per well, respectively, in 90 μl/well volumes in a 96-well plate. Cells were allowed to attach and equilibrate overnight at 37°C, 5% CO_2. Serial dilutions of lactacystin were prepared in culture medium, and 10 μl each dilution was added to wells. Cells were incubated at 37°C, 5% CO_2 for 105 min. The plate was removed and allowed to equilibrate to 22°C before 100 μl/well of Proteasome-Glo™ Cell-Based Reagent was added. Luminescence was measured with a DYNEX MLX® luminometer 15 min after adding reagent.

chymotrypsin-like, trypsin-like, and caspase-like, clearly indicating that other proteases have similar catalytic properties. By extension, designing peptide substrates that are unique for proteasome activity is difficult. To overcome this problem, we developed a permeabilizing agent that enhances access to the proteasome while leaving lysosomal vesicles generally intact, and thus minimizing the effects of non-specific proteases. This method proved very amenable for developing a direct, cellular assay for the chymotrypsin-like activity of the proteasome using the Suc-LLVY-aminoluciferin substrate. Numerous cell lines have been tested with the assay, including Jurkat, U937, U266, H929, RPMI-8226, HL-60, H226, PA-1, DU 145, and MCF-7. The specificity of the assay can be confirmed by inhibiting a majority of the activity with the selective proteasome inhibitor, lactacystin *(24–26)* (*see* **Notes 2** and **3**). **Figure 4** shows examples for two cell lines. The cell-based proteasome assay does not require lysate preparation and enables measurement of proteasome activity in a more physiologically relevant environment. Achieving specificity for proteasome activity in a direct cell-based assay using bioluminescent substrates for the trypsin-like and caspase-like activities is more challenging and is currently being investigated.

2. Materials

2.1. Equipment

1. White-walled multiwell plates. Solid bottom plates are optimal for enzyme assays, and clear bottom plates are optimal for cellular assays.
2. Multichannel pipette or automated pipetting station for delivery of Proteasome-Glo™ Reagent.
3. Plate shaker for mixing multiwell plates.
4. Luminometer capable of reading multiwell plates (Glo-Max™ Microplate Luminometer).
5. Fluorimeter capable of reading multiwell plates (LabSystems Fluoroskan Ascent).
6. 37°C incubator with 5% CO_2.

2.2. In Vitro Assays

1. Proteasome-Glo™ Chymotrypsin-Like Assay, Proteasome-Glo™ Trypsin-Like Assay, and Proteasome-Glo™ Caspase-Like Assay (Promega, Madison, WI).
2. 20S proteasome enzyme (Biomol, Plymouth Meeting, PA, or Boston Biochem, Cambridge, MA) or 26S proteasome (Biomol).
3. 10 mM HEPES buffer, pH 7.6 (for proteasome dilution).
4. The proteasome inhibitors, lactacystin, clasto-lactacystin-β-lactone, and epoxomicin (Biomol).
5. Dimethylsulfoxide (DMSO) (Hybri-Max®, Sigma, St. Louis, MO).

6. Suc-LLVY-AMC (Calbiochem, EMD Biosciences, Ca Jolla, CA, USA).
7. Boc-LRR-AMC (Biomol).
8. Ac-nLPnLD-AMC (Biomol).

3. Methods

The methods described below outline (i) in vitro (biochemical) assays for monitoring all three catalytic activities of the proteasome using purified 20S proteasome, (ii) a direct, cellular assay for monitoring the chymotrypsin-like activity of the proteasome, (iii) inhibition studies using both formats, and (iv) a protocol for multiplexing the bioluminescent, cellular proteasome assay and a fluorescent assay for caspase activity.

3.1. In Vitro Proteasome Assays

Directions are given for performing the in vitro Proteasome-Glo™ Assays in a total volume of 100 μl using 96-well plates and a luminometer. However, the assays can be easily adapted to different volumes providing the 1:1 ratio of Proteasome-Glo™ Reagent volume to sample volume is preserved (e.g., 25 μl sample + 25 μl Proteasome-Glo™ Reagent in a 384-well format).

3.1.1. Proteasome-Glo™ Reagent Preparation

1. Thaw the Proteasome-Glo™ Buffer and equilibrate both buffer and the lyophilized Luciferin Detection Reagent to room temperature prior to use.
2. Reconstitute the Luciferin Detection Reagent in the amber bottle by adding the appropriate volume of Proteasome-Glo™ Buffer. The Luciferin Detection Reagent should go into solution easily in less than 1 min.
3. Thaw the appropriate substrate and equilibrate to room temperature prior to use. For the Chymotrypsin-like Assay, use the Suc-LLVY-Glo™ Substrate; for the Trypsin-like Assay, use the Z-LRR-Glo™ Substrate; and for the Caspase-like Assay, use the Z-nLPnLD-Glo™ Substrate. A slight precipitate may be observed. Mix well by vortexing briefly.
4. Prepare the Proteasome-Glo™ Reagent by adding the Proteasome-Glo™ Substrate to the resuspended Luciferin Detection Reagent as per **Table 1**. Label the reagent bottle to identify the substrate used.
5. Allow the Proteasome-Glo™ Reagent to sit at room temperature for 30 min prior to use. This allows for the removal of any contaminating free aminoluciferin. Although free aminoluciferin is not detected by HPLC, it is present in trace amounts (*see* **Fig. 5**).

Table 1
Instruction for making the Proteasome-Glo™ Reagents.

Proteasome-Glo™ Assay	Cat. no.	Substrate	Volume substrate added (µl)	Substrate concentration in reagent (µM)
Chymotrypsin-like assay	G8621	Suc-LLVY-Glo™	50	40
Chymotrypsin-like assay	G8622	Suc-LLVY-Glo™	250	40
Trypsin-like assay	G8631	Z-LRR-Glo™	100	30
Trypsin-like assay	G8632	Z-LRR-Glo™	500	30
Caspase-like assay	G8641	Z-nLPnLD-Glo™	50	40
Caspase-like assay	G8642	Z-nLPnLD-Glo™	250	40

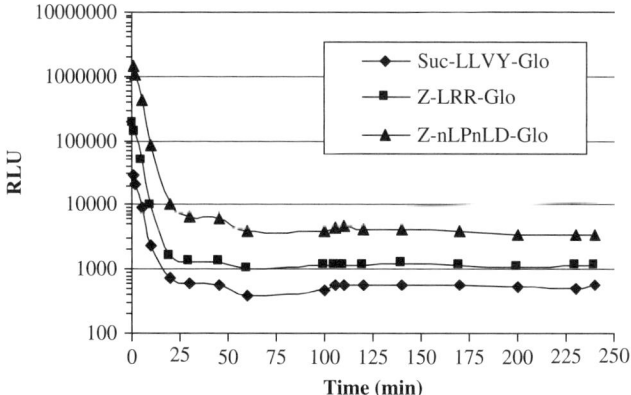

Fig. 5. Time course of free aminoluciferin removal from the Proteasome-Glo™ Reagent. The proteasome substrates (Suc-LLVY-Glo™, Z-LRR-Glo™, and Z-nLPnLD-Glo™) were added to the reconstituted Luciferin Detection Reagent, and a time course of luminescence loss was recorded. Trace amounts of free aminoluciferin are present in the substrate and are removed by incubation with the reconstituted Luciferin Detection Reagent. To achieve maximal assay sensitivity with minimal background luminescence, the prepared Proteasome-Glo™ Reagent should be incubated for 30 min before use.

3.1.2. Proteasome-Glo™ Assay Conditions

Prepare the following reactions to detect proteasome activity (or inhibition of activity) in purified enzyme preparations:

- **Blank:** Proteasome-Glo™ Reagent + vehicle control for test compound or inhibitor, if used.
- **Positive Control:** Proteasome-Glo™ Reagent+vehicle control+purified proteasome enzyme (20S or 26S).
- **Assay:** Proteasome-Glo™ Reagent + test compound + purified proteasome enzyme (20S or 26S).

The blank is used as a measure of any background luminescence associated with the test compound vehicle and the Proteasome-Glo™ Reagent and should be subtracted from experimental values. The positive control is used to determine the maximum luminescence obtainable with the purified enzyme system. Vehicle refers to the solvent used to dissolve the inhibitor or test compound used in the study.

3.1.3. Proteasome-Glo™ Standard Assay (96-well, 100 µl Final Reaction Volume)

1. Add 50 µl Proteasome-Glo™ Reagent to each well of a white 96-well plate containing 50 µl blank, control, or test sample. If reusing tips, be careful not to touch pipette tips to the wells containing samples to avoid cross-contamination.
2. Gently mix contents of wells using a plate shaker at (*see* **Notes 4–6**). for 30 s. Incubate at room temperature for 10 min to 3 h depending on convenience of reading time. Maximal signal is reached typically within 10–30 min using purified 20S proteasome (*see* **Fig. 2**). At this time, sensitivity is optimal. Temperature fluctuations will impact the luminescent readings; if the room temperature fluctuates too much, a constant-temperature incubator may be desired.
3. Record luminescence with a plate-reading luminometer as directed by the manufacturer.

3.1.4. Determining Inhibition Curves for the Three Proteasome Catalytic Activities

1. When generating IC_{50} curves for a proteasome inhibitor, titrate the inhibitor in HEPES (10 mM, pH 7.6) and add 25 ul per well in a 96-well plate. Dilute 20S or 26S proteasome in the same buffer to 2 µg/ml and add 25 µl per well for a final proteasome concentration of 1 µg/ml.
2. Incubate at room temperature for 1 hour to allow irreversible inhibitors such as lactacystin and epoxomicin to bind completely.
3. Add 50 µl Proteasome-Glo™ Reagent, containing Suc-LLVY-Glo™, Z-LRR-Glo™, or Z-nLPnLD-Glo™, to each well of a white 96-well plate.

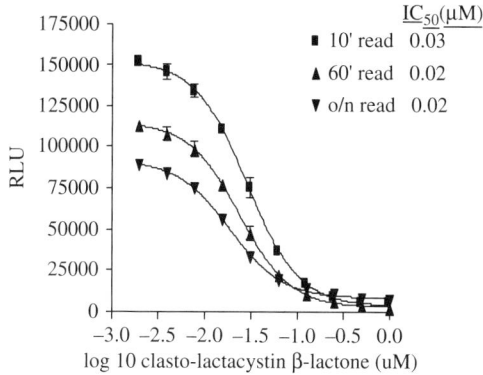

Fig. 6. Inhibition of proteasome with clasto-lactacystin β lactone. Inhibitor titrations and 26S proteasome were combined in 96-well plates as described in Section 3.1.4. All three proteasome activities were tested with clasto-lactacystin β lactone, the active analog of lactacystin *(26)*. Inhibition curves using the Proteasome-Glo™ Chymotrypsin-Like Assay demonstrate consistent IC_{50} values when readings are taken between 10 min and 18 h after addition of the Proteasome-Glo™ Reagent. Within 10 min, the dynamic range is maximal.

4. Gently mix contents of wells using a plate shaker for 30 s. Incubate at room temperature for at least 10 min.
5. Record luminescence with a plate-reading luminometer. Luminescence can be read at various times (*see* **Fig. 6** and **Table 2**).

Table 2
A Summary of the IC_{50} values for clasto-lactacystin β lactone on all three proteasome activities. IC_{50} values were calculated from readings taken at 60 min. The relative potencies are consistent with previous reports *(23,24)*.

	Inhibitor IC_{50} (μM)
Substrate	Clasto-lactacystin-β-lactone
Suc-LLVY-Glo™	0.02
Z-LRR-Glo™	0.76
Z-nLPnLD-Glo™	2.6

3.2. Detection of Proteasome Activity from Cultured Cells

This protocol provides instructions for performing the Proteasome-Glo™ Cell-Based Assay in a total volume of 200 μl using 96-well plates and a luminometer. However, the assay can be easily adapted to different volumes if the 1:1 ratio of Proteasome-Glo™ Cell-Based Reagent volume to sample volume is preserved (e.g., 25 μl sample + 25 μl Proteasome-Glo™ Cell-Based Reagent in a 384-well format).

3.2.1. Proteasome-Glo™ Cell-Based Reagent Preparation

1. Thaw the Proteasome-Glo™ Cell-Based Buffer, and equilibrate both the buffer and the lyophilized Luciferin Detection Reagent to room temperature before use.
2. Reconstitute the Luciferin Detection Reagent in the amber bottle by adding the appropriate volume of Proteasome-Glo™ Cell-Based Assay Buffer (10 ml each for cat. no. G8660 or G8661 and 50 ml for cat. no. G8662). The Luciferin Detection Reagent should go into solution easily in less than 1 min.
3. Thaw the Suc-LLVY-Glo™ Substrate and equilibrate to room temperature before use. A slight precipitate may be observed. Mix well by vortexing briefly.
4. Prepare the Proteasome-Glo™ Cell-Based Reagent by adding the Suc-LLVY-Glo™ Substrate to the resuspended Luciferin Detection Reagent. For cat. no. G8660 and G8661, add 50 μl Suc-LLVY-Glo™ Substrate to the 10 ml Luciferin Detection Reagent. For cat. no. G8662, add 250 μl Suc-LLVY-Glo™ Substrate to the 50 ml Luciferin Detection Reagent. Mix to homogeneity by swirling the contents or inverting the bottle. The Suc-LLVY-Glo™ Substrate will be at a concentration of 40 μM in the Proteasome-Glo™ Cell-Based Reagent and 20 μM in the final assay. The apparent K_m for the substrate is approximately 60 μM using cells.
5. Allow the Proteasome-Glo™ Cell-Based Reagent to stand at room temperature for 30 min before use (*see* **Subheading 3.1.1.** and **Fig. 5**).

3.2.2. Controls and Assay Conditions

Prepare the following reactions to detect proteasome activity (or inhibition of activity) using cells in culture:

- **Blank:** Proteasome-Glo™ Cell-Based Reagent + culture medium (without cells) and vehicle control used.
- **No-Treatment Control:** Proteasome-Glo™ Cell-Based Reagent + culture medium containing cells and vehicle control (without test compound).
- **Inhibitor Control:** Proteasome-Glo™ Cell-Based Reagent + culture medium containing cells with a specific proteasome inhibitor such as lactacystin or epoxomicin.

- **Test:** Proteasome-Glo™ Cell-Based Reagent + culture medium containing cells with test compound.

The **blank** is used as a measure of background luminescence contributed by the cell-culture medium, the vehicle used to deliver test compounds, and the Proteasome-Glo™ Cell-Based Reagent and should be subtracted from all control and assay values. Vehicle refers to the solvent used to dissolve the inhibitor or test sample used in the study. The **no-treatment control** is used to determine the maximum luminescence obtained from untreated cells. The **inhibitor control** is used to determine the maximum inhibition of proteasome activity and helps identify non-specific protease activity not related to the proteasome. **Test** samples represent the cells with their respective treatments.

3.2.3. Proteasome-Glo™ Cell-Based Standard Assay (96-Well Plates)

1. Prepare the Proteasome-Glo™ Cell-Based Reagent as described in Section 3.2.1 and mix thoroughly before starting the assay.
2. Optimize cell number and treatment duration for each cell line. For a 96-well plate format, we recommend working with approximately 10,000–20,000 suspension cells per well or 5000–10,000 adherent cells per well (*see* **Note 7**).
3. For consistent results, equilibrate assay plates to a constant temperature before performing the assay (*see* **Note 8**).
4. Use identical cell numbers and volumes for the assay and control reactions.
5. If preparing multiple plates, controls must be replicated on each plate.
6. Add 100 µl Proteasome-Glo™ Cell-Based Reagent to each 100 µl sample and appropriate controls as needed. Cover the plate with a plate sealer or lid.
7. Mix the contents of the wells using a plate shaker for 2 min. Incubate at room temperature for a minimum of 10 min.
8. Measure the luminescence of each sample in a plate-reading luminometer as directed by the manufacturer.

3.2.4. Determining Inhibition of the Proteasome in Cultured Cells

1. Add cells to 96-well plates in a 90 µl volume at 5000–20,000 cells per well (*see* **Note 4**) and allow the cells to equilibrate for 2 h (suspension cells) or overnight (attached cells) at 37°C, 5% CO_2.
2. Serial dilute the inhibitor in culture medium and add 10 µl per well, including a no inhibitor control.
3. Incubate the inhibitor with the cells for 1–2 h.
4. Add the Proteasome-Glo™ Cell-Based Reagent, shake the plate, and measure the luminescence as in Section 3.2.3 (*see* **Fig. 7**).

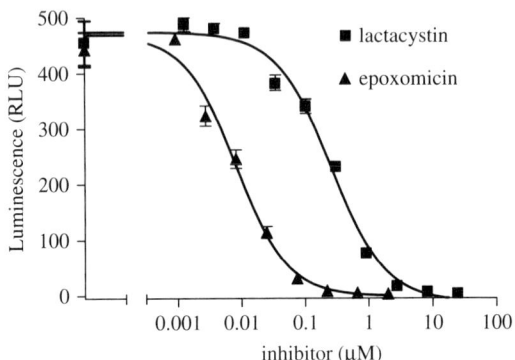

Fig. 7. Comparison of two proteasome inhibitors tested in cultured cells. U266 multiple myeloma cells were grown in RPMI-1640 containing 10% FBS and 1 mM sodium pyruvate. Cells were added to 96-well plates 10,000 cells per well in 90 µl per well. Cells were then equilibrated at 37°C, 5% CO_2 for 2 h. Serial dilutions of lactacystin or epoxomicin were prepared in culture medium, and 10 µl each dilution was added to wells. The cells were incubated with the drugs for 105 min at 37°C, 5% CO_2. The plate was allowed to equilibrate to 22°C before 100 µl per well of Proteasome-Glo™ Cell-Based Reagent was added. Luminescence was measured with a DYNEX MLX® luminometer 15 min after adding reagent. The relative potency for the two inhibitors is consistent with published information *(27)*.

3.2.5. Multiplexing the Proteasome-Glo™ Cell-Based Assay with a Caspase Activity Assay

1. Add cells in culture medium to 96-well plates in 90 µl per well and allow to equilibrate overnight at 37°C, 5% CO_2.
2. Add inhibitor or test drug to cells in 10 µl per well and incubate for various times at 37°C, 5% CO_2.
3. Remove plates from the incubator and allow them to equilibrate to room temperature.
4. Prepare the Proteasome-Glo™ Cell-Based Reagent as in Section 3.2.1.
5. Prepare a modified Apo-ONE® Reagent by adding the Apo-ONE® substrate 1:20 into the Apo-ONE® Buffer. This gives a 10× Apo-ONE® Homogeneous Caspase 3/7 Reagent.
6. Add the Proteasome-Glo™ Cell-Based Reagent to the cells at 100 µl per well.
7. Incubate for 15 min and measure luminescence on a plate-reading luminometer.
8. Add 20 ul per well of the 10× Apo-ONE® Reagent.
9. Shake the plates for 2 min and incubate at room temperature for 30 min.
10. Measure fluorescence at 485/527 nm (*see* **Fig. 8**).

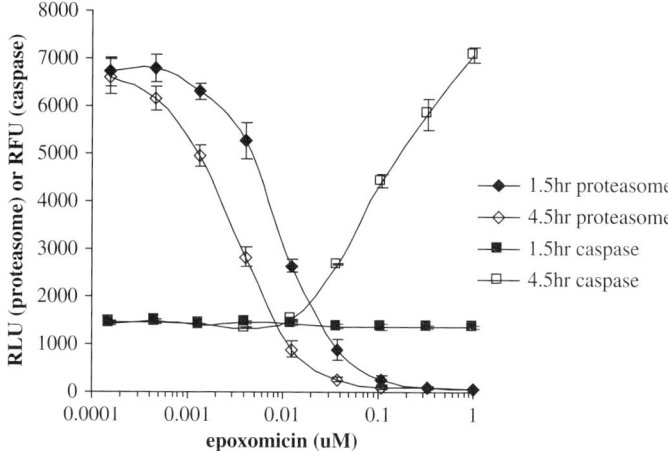

Fig. 8. Sequential multiplex to determine proteasome and caspase-3/7 activity. H929 multiple myeloma cells were grown in RPMI-1640 containing 10% FBS and 1 mM sodium pyruvate; they were added to 96-well plates 10,0000 cells per well in 90 µl per well and incubated overnight at 37°C, 5% CO_2. Epoxomicin was titrated in culture medium and added at 10 µl per well. Cells were incubated with the inhibitor for 1.5 or 4.5 h at 37°C, 5% CO_2. The plate was removed from the incubator and equilibrated to room temperature before adding Proteasome-Glo™ Cell-Based Reagent. The plate was mixed, and luminescence was recorded after 15 min. A 10× Apo-ONE® Homogeneous Caspase 3/7 Reagent containing a $(Z - DEVD)_2$-Rhodamine 110 was then added at 20 µl per well, and fluorescence was recorded after 30 min. Epoxomicin treatment for 1.5 h did not induce caspase-3/7 activity, but caspase-3/7 activity was induced after 4.5 h treatment with >0.04 µ M epoxomicin.

4. Notes

1. SDS cannot be used as an activating agent for the bioluminescent assay. Although SDS is frequently used to enhance the chymotrypsin-like activity of the proteasome, it is detrimental to luciferase and is not necessary for these bioluminescent assays. Superior sensitivity is achieved even in the absence of SDS (*see* **Fig. 3**).
2. The Proteasome-Glo™ 3-Substrate System is optimized for use with purified 20S or 26S proteasome. It may be possible to assay for proteasome activity in crude cell lysates using the luminogenic substrates, but controls for specificity should be included. None of the substrates are uniquely cleaved by the proteasome; therefore, depending on the extraction method and level of purity of the proteasome, confirming specificity with specific inhibitors may be critical. Lactacystin and epoxomicin are natural inhibitors that are very selective for the proteasome *(27,28)*, although lactacystin has been reported to inhibit cathepsin A and tripeptidyl

peptidase II under certain circumstances *(28,29)*. Epoxomicin and lactacystin are the most selective commercially available inhibitors and are most potent against the chymotrypsin-like activity, followed by the trypsin-like and caspase-like activities (*see* table in **Fig. 6**) *(28)*. Demonstrating specificity for the trypsin-like and caspase-like activities of the proteasome is more difficult because of the lack of commercially available potent and selective inhibitors.

3. The Proteasome-Glo™ Cell-Based Assay is formulated to minimize non-proteasome cleavage of the Suc-LLVY-Glo™ Substrate. However, some cell lines may contain protease activity that cannot be inhibited using either lactacystin or epoxomicin. We recommend performing a proteasome-inhibitor control as well as an untreated cell control in each assay plate to help define this window of activity attributable to the chymotrypsin-like activity of the proteasome. Uninhibitable activity is typically low and can be subtracted (*see* **Fig. 7**).

4. The chemical environment of the luciferase reaction will affect the enzymatic rate and thus luminescence intensity. Solvents used for various chemical compounds may interfere with the luciferase reaction and thus the light output from the assay. DMSO, commonly used as a vehicle to solubilize organic chemicals, has been tested at final concentrations up to 1% in the assay and found to have a minimal effect on light output. Libraries stored in DMSO are compatible with the bioluminescent proteasome assays.

5. Owing to the sensitivity of the Proteasome-Glo™ Assays, contamination with other luciferin-containing reagents can result in high background luminescence. Be sure that shared luminometers are cleaned thoroughly before performing this assay. Avoid workspaces and pipettes that are used with luciferin-containing solutions, including luminescence-based cell viability, apoptosis, or gene reporter assays.

6. The final concentration of 20S proteasome should be within the linear range of the assay (*see* **Fig. 3**). With the enhanced sensitivity of the bioluminescent assays, less 20S proteasome is typically needed for the assays. We recommend defining the linear range for the particular proteasome preparation.

7. Owing to the sensitivity of these assays, researchers are encouraged to use an appropriate number of cells to stay within the linear range for the Proteasome-Glo™ Cell-Based Assay. For a 96-well plate format, we recommend preparing bioassays to contain approximately 10,000–20,000 suspension cells per well or 5000–10,000 adherent cells per well. Cell number can be scaled accordingly when using smaller formats. Empirical determination of the optimal cell number and treatment duration for each cell line and plate format may allow the use of even fewer cells; proteasome activity may vary significantly depending on cell type.

8. Environmental factors that affect the rate of the luciferase reaction will also affect the intensity of the light output and the stability of the luminescent signal. Temperature can affect the rate of this enzymatic assay and thus the light output. For consistent results, equilibrate assay plates to a constant temperature before performing the

assay. For batch-mode processing of multiple plates, positive and negative controls should be included for each plate. Additionally, precautions should be taken to ensure complete temperature equilibration.

Acknowledgments

The authors thank colleagues at Promega Biosciences, Michael Scurria, Laurent Bernad, Bill Dailey, and James Unch, for synthesizing the bioluminescent proteasome substrates. We are indebted to Keith Wood and Dieter Klaubert for the homogeneous, bioluminescent assay concept. We also thank Kay Rashka, Sandra Hagen, Jeri Culp, Debra Lange, Brian McNamara, Anissa Moraes, and Pam Guthmiller for translating the concepts into products.

References

1. Adams, J., Palombella, V.J., Sausville, E.A., Johnson, J., et al. (1999) Proteasome inhibitors: a novel class of potent and effective antitumor agents. *Cancer Res.* **59**, 2615–2622.
2. Adams, J. (2002) Development of the proteasome inhibitor PS-341. *Oncologist* **7**, 9–16.
3. Voorhees, P.M., Dees, E.C., O'Neil, B., and Orlowski, R.Z. (2003) The proteasome as a target for cancer therapy. *Clin. Cancer Res.* **9**, 6316–6325.
4. Burger, A.M., and Seth, A.K. (2004) The ubiquitin-mediated protein degradation pathway in cancer: therapeutic implications. *Eur. J. Cancer* **4**, 2217–2229.
5. Gu, Z., Nakamura, D., Yao, D., Shi, Z.-Q., and Lipton, S.A. (2005) Nitrosative and oxidative stress links dysfunctional ubiquitination to Parkinson's disease. *Cell Death Diff.* **12**, 1202–1204.
6. Baumeister, W., Walz, J., Zühl, F., and Seemüller, E. (1998) The proteasome: paradigm of a self-compartmentalizing protease. *Cell* **92**, 367–380.
7. Wolf, D.H., and Hilt, W. (2004) The proteasome: a proteolytic nanomachine of cell regulation and waste disposal. *Biochem. Biophys. Acta* **1695**, 19–31.
8. Glickman, M.H., and Ciechanover, A. (2002) The ubiquitin-proteasome proteolytic pathway: destruction for the sake of construction. *Physiol. Rev.* **82**, 373–428.
9. Rajkumar, S.V., Richardson, P.G., Hideshima, T., and Anderson, K.C. (2005) Proteasome inhibition as a novel therapeutic target in human cancer. *J. Clin. Oncol.* **23**, 630–639.
10. Nussbaum, A.K., Dick, T.P., Keilholz, W., Schirle, M., et al. (1998) Cleavage motifs of the yeast 20S proteasome subunits deduced from digest of enolase I. *Proc. Natl. Acad. Sci. U. S. A.* **95**, 12504–12509.
11. Rechsteiner, M., and Hill, C.P. (2005) Mobilizing the proteolytic machine: cell biological roles of proteasome activators and inhibitors. *Trends Cell Biol.* **15**, 27–33.

12. Kisselev, A.F., Kaganovich, D., and Goldberg A.L. (2002) Binding of hydrophobic peptides to several non-catalytic sites promotes peptide hydrolysis by all active sites of 20S proteasomes. *J. Biol. Chem.* **277**, 22260–22270.

13. Kisselev, A.F., Garcia-Calvo, M., Overkleeft, H.S., Peterson, E., et al. (2003) The caspase-like sites of proteasomes, their substrate specificity, new inhibitors and substrates, and allosteric interactions with the trypsin-like sites. *J. Biol. Chem.* **278**, 35869–35877.

14. Ciechanover, A. (2005) Intracellular protein degradation: from a vague idea thru the lysosome and the ubiquitin-proteasome system and onto human diseases and drug targeting. *Cell Death Diff.* **12**, 1178–1190.

15. Chauhan, D., Hideshima, T., and Anderson, K.C. (2005) Proteasome inhibition in multiple myeloma: therapeutic implication. *Ann. Rev. Pharmacol. Toxicol.* **45**, 465–476.

16. Chauhan, D., Catley, L., Li, G., Podar, K., et al. (2005) A novel orally active proteasome inhibitor induces apoptosis in multiple myeloma cells with mechanisms distinct from Bortezomib. *Cancer Cell* **8**, 407–419.

17. Richardson, P.G., Barlogie, B., Berenson, J., Singhal, S., et al. (2003) Phase 2 study of bortezomib in relapsed, refractory myeloma. *N. Engl. J. Med.* **348**, 2609–2617.

18. Papandreou, C.N., and Logothetis, C.J. (2004) Bortezomib as a potential treatment for prostate cancer. *Cancer Res.* **64**, 5036–5043.

19. Leytus, S.P., Melhado, L.L., and Mangel, W.F. (1983) Rhodamine-based compounds as fluorogenic substrates for serine proteinases. *Biochem. J.* **209**, 299–307.

20. Liu, J., Bhalgat, M., Zhang, C., Diwu, Z., Hoyland, B., and Klaubert, D.H. (1999) Fluorescent molecular probes V: a sensitive caspase-3 substrate for fluorometric assays. *Bioorg. Med. Chem. Lett.* **9**, 3231–3236.

21. Grant, S.K., Sklar, J.G., and Cummings, R.T. (2002) Development of novel assays for proteolytic enzymes using rhodamine-based fluorogenic substrates. *J. Biomol. Screen.* **7**, 531–540.

22. O'Brien, M.A., Daily, W.J., Hesselberth, P.E., Moravec, R.A., et al. (2005). Homogeneous, bioluminescent protease assays: caspase-3 as a model. *J. Biomol. Screen.* **10**, 137–148.

23. Lightcap, E.S., Mccormack, T.A., Pien, C.S., Chau, V., et al. (2000) Proteasome inhibition measurements: clinical application. *Clin. Chem.* **46**, 673–683.

24. Fenteany, G., Standaert, R.F., Lane, W.S., Choi, S., et al. (1995) Inhibition of proteasome activities and subunit-specific amino-terminal threonine modification by lactacystin. *Science* **268**, 726–731.

25. Dick, L., Cruikshank, A., Grenier, L., Melandri, F.D., et al. (1996) Mechanistic studies on the inactivation of the proteasome by lactacystin: a central role for clasto-lactacystin β-lactone. *J. Biol. Chem.* **271**, 7273–7276.

26. Dick, L., Cruikshank, A., Destree, A.T., Grenier, L., et al. (1997) Mechanistic studies on the inactivation of the proteasome by lactacystin in cultured cells. *J. Biol. Chem.* **272**, 182–188.
27. Meng, L., Mohan, R.L., Kwok, B.H.B., Elofsson, M., Sin, N., and Crews, C.M. (1999) Epoxomicin, a potent and selective proteasome inhibitor, exhibits in vivo anti-inflammatory activity. *Proc. Natl. Acad. Sci. U. S. A.* **96**, 10403–10408.
28. Corey, E.J., and Li, W.-D.Z. (1999) Total synthesis and biological activity of lactacystin, omuralike and analogs. *Chem. Pharm. Bull.* **47**, 1–10.
29. Wojcik, C., and Napoli, M.D. (2004) Ubiquitin-proteasome system and proteasome inhibition: new strategies in stroke therapy. *Stroke* **35**, 1506–1518.

14

A Simple Method for Profiling miRNA Expression

Jia-Wang Wang and Jin Q. Cheng

Summary

Here we describe a simple protocol that uses positively charged nylon membrane dot blot to profile miRNA expression. A library of 515 antisense oligodeoxynucleotides of human and mouse mature miRNAs was synthesized and spotted on GeneScreen Plus membrane using a dot-blot equipment. Total RNA or enriched small molecular weight RNAs (smwRNAs) were enzymatically radiolabeled by poly (A) polymerase and then hybridized to the nylon membrane oligo arrays. The spot signal intensity on the membrane was analyzed using phosphorimaging. This method offers a convenient and economic way to simultaneously detect the expression of hundreds of miRNAs.

Key Words: Dot blot; microRNA; hybridization; oligodeoxynucleotide library; phosphorimaging.

1. Introduction

MicroRNAs are a class of small noncoding RNAs that are highly evolutionarily conserved and have been shown to negatively regulate the expression of targeted gene(s). miRNAs have been identified in almost all metazoans, ranging from plants to worms, flies, and human (*1*). To date, more than 515 human miRNAs have been identified. It has been predicted that the number of miRNAs could exceed 1000 (*2,3*). Most miRNAs are present at very high steady-state levels from 1000 molecules to 50,000 molecules per cell (*4*). Like conventional, protein-coding mRNA, miRNAs are transcribed by RNA polymerase II, spliced and polyadenylated (called primitive miRNA or pri-miRNA) (*5*). However, unlike mRNA, the pri-miRNAs contain a stem-loop structure that

From: *Methods in Molecular Biology, vol. 414: Apoptosis and Cancer*
Edited by: G. Mor and A. B. Alvero © Humana Press Inc., Totowa, NJ

can be recognized and excised by the RNAi machinery to generate hairpin "precursor" miRNAs (pre-miRNA) that are approximately 70 nucleotides in animals or approximately 100 nucleotides in plants. Pre-miRNAs are exported to the cytoplasm for further procession that results in a "mature" miRNA of approximately 21–25 nucleotides, which has a 5' phosphate and free 3' hydroxyl *(5,6)*.

The mature miRNA then guides a complex called miRNA-containing ribonucleo-protein particles (miRNP) to the complementary site(s) in the 3'untranslated region (UTR) of a target mRNA. Consequently, translation blockade or mRNA degradation will occur depending on whether it is partially matched or completely matched, respectively *(7)*. In plants, miRNAs bind to a single, general site in either the coding or 3' UTRs of the target mRNA; in animals, miRNAs bind to multiple, partially complementary sites in the 3' UTRs *(8)*. Bioinformatics analyses suggest that as much as 30% of all human genes may be under miRNA regulation *(9)*. Moreover, the levels of individual miRNAs are dramatically changed in different cell types and different developmental stages *(4)*, suggesting that miRNA may play a role in fundamental processes such as cell differentiation, communication, and death during development. Thus, misregulation of miRNA function might contribute to human diseases like tumorigenesis and infectious diseases *(10–12)*. Therefore, it is not surprising that there is an explosion of interest in miRNAs from researchers in all areas of cell and developmental biology, and the miRNA expression profiling would be the first step to study the biological functions and gene regulation of miRNAs.

The DNA microarray technology is a powerful tool to profile mRNA expression. Predictably, it has been successfully applied to miRNA profiling, and several techniques have been developed recently *(13–16)*. However, chip microarray-based miRNA techniques require expensive equipments and reagents, and their uses are limited in a very small number of laboratories. In contrast, because there are just hundreds of miRNA for each species, the membrane-based profiling is quite feasible and could be widely used as a routine technique. Here, we describe a method developed in our laboratory that allows economically and sensitively detect miRNA expression, which has also been used by other laboratories *(17)*. We use positive-charged nylon membrane to make dot-blot oligo array to detect miRNA expression. First, a library of antisense oligodeoxynucleotides of mature miRNAs was synthesized and spotted on GeneScreen Plus hybridization transfer membrane using a Bio-Dot equipment. Then, total RNA or enriched small molecular RNA will be directly [^{32}P]-radiolabeled using yeast poly (A) polymerase. Finally, the nylon membrane oligo arrays were used for hybridization with the P^{32}-labeled probe, and the spot signal intensity on the membrane will be analyzed using phosphorimaging and/or autoradiography.

2. Materials

2.1. Equipments

1. Bio-Dot Microfiltration Apparatus (cat. no. 170-6545, Bio-Rad, Hercules, CA, USA).
2. Stratalinker® 2400 UV Crosslinker (Stratagene, La Jolla, CA, USA).
3. Microcon YM-100 (Millipore).
4. HB1D hybridization tubes (44 mm × 200 mm).
5. Hybridization incubator (cat. no. HB-1D, Techne, Cambridge, UK).
6. Phosphor storage screen.
7. Typhoon™, Phosphor Imager Imaging systems (GE Healthcare Life Sciences, Pittsburgh, PA, USA).
8. Image Eraser (GE Healthcare Life Sciences).

2.2. Reagents

1. miRNA oligonucleotide synthesis: miRNA mature sequences were obtained from miRBase (http://microrna.sanger.ac.uk/), converted to complementary DNA sequence and triplicate the sequences, then chemically synthesized [also include several positive controls (tRNA) and negative controls] at Integrated DNA Technologies (IDT, Coralville, IA, USA) at 25 nmol scale and diluted at 50 μm in a 96-well plate format.
2. GeneScreen Plus Hybridization Transfer Membrane (cat. no. NEF988, PerkinElmer, Waltham, MA, USA).
3. Wetting buffer: 0.4 M Tris–HCl, pH 7.5.
4. Washing buffer: 0.5 N NaCl, 0.5 M Tris–HCl, pH 7.5.
5. TE buffer: 10 mM Tris–Cl, pH 7.5, 1 mM EDTA.
6. 2× denature buffer: 0.5 N NaOH, 1 M NaCl.
7. Dilution buffer: 0.1× SSC, 0.125 N NaOH.
8. G25 column filtration (cat. no. PD10, Amersham Pharmacia Biotech).
9. TRIZOL® Reagent (cat. no. 15596-026).
10. Chloroform.
11. Isopropyl alcohol.
12. 75% ethanol (in DEPC-treated water).
13. RNase-free water.
14. 1 M Tris (1000 ml): 121.1 g Tris, 800 ml H_2O, 60 ml HCl to pH 7.5. Autoclave.
15. 10 N NaOH (500 ml): 200 g in 500 ml.
16. 20× SSC (1000 ml): 175.3 g NaCl + 88.2 g sodium citrate + 800 ml ddH_2O, pH 7.0 by NaOH. Add ddH2O to reach final volume of 1000 ml. Autoclave.

17. 5 M NaCl (1000 ml): 292.2 g. Autoclave.
18. ^{32}P-α-ATP (Amersham Pharmacia Biotech).
19. RNasin ribonuclease inhibitor (Promega, Madison, WI, USA), stored at –20°C.
20. Yeast poly (A) polymerase (USB, Cleveland, OH, USA), stored at –20°C.
21. MicroHyb Hybridization Buffer (Invitrogen, Carlsbad, CA, USA), stored at room temperature. If crystals formed, dissolve at 42°C water bath before use.

3. Methods

The methods described in this section include (i) preparation of miRNA oligo dot-blot nylon membrane array, (ii) miRNA extraction and enrichment from tissues and cell cultures, (iii) RNA labeling by poly (A) polymerase, and (iv) hybridization and detection by phosphorimaging.

3.1. Preparation of miRNA Oligo Dot-Blot Nylon Membrane Array

We use nylon membrane (dot-blot) arrays spotted with oligodeoxynucleotides antisense to mature miRNAs. miRNAs from tissues or cells are radiolabeled and used as probes to hybridize with the oligos on the membrane. The spot signal intensity on the membrane was quantitated using a phosphorimager.

1. Cut GeneScreen Plus Hybridization Transfer Membrane (cat. no. NEF988, PerkinElmer) to the exact size (95 mm × 120 mm) of the dot manifold (cat. no. 170-6545, Bio-Dot Microfiltration Apparatus, Bio-Rad, Hercules, CA, USA).
2. Wet the membrane in wetting buffer and soak for 5 min.
3. Place the wet membrane on the manifold and assemble according to manufacturers' recommendations. The manifold should be well cleaned and rinsed at least three times with Millipore water prior to use.
4. Dilute oligo in TE to 10 µM in a 96-well plate to make duplicate dot blots using multiple channel pipettor (50 µM oligo stock 4 µl, TE buffer 16 µl).
5. Adding equal volume (20 µl) of 2× denature buffer to the diluted oligos, mix well, and denature DNA for 10 min at room temperature (10 µM diluted oligo 20 µl, 2× denature buffer 20 µl).
6. Add 40 µl H$_2$O and dilute DNA samples on ice to desired concentrations in 150 µl dilution buffer to make a volume of total 230 µl.
7. Load 100 µl samples into the manifold.
8. Apply a light suction to the manifold until the loading buffer is drawn through the wells, approximately 30 s.
9. Add 100 µl washing buffer to each well and apply a light suction as in step 8 to neutralize and wash the membrane.
10. Remove the membrane from the manifold and place the wet membrane on a piece of wet filter paper. This will prevent the membrane from drying.
11. Fix the DNA to the membrane by UV Autocross-linking (1200 uJ × 100, Stratalinker® 2400 UV Crosslinker, 120 V).
12. Repeat steps 8–11 for another blot.

13. Proceed directly to the prehybridization step or air-dry the membranes and store them at 4°C in a plastic bag for future use.

3.2. miRNA Extraction and Enrichment from Tissues and Cell Cultures

The starting material for detection of miRNAs can be either total RNA or enriched small molecular weight RNA (smwRNA). Total RNA is isolated by standard TRIZOL (guanidinium isothiocyanate/acidic phenol) method according to manufacturers' instructions with the following modifications (*see* **Note 1**). smwRNAs are enriched by Microcon YM-100.

1. Grow cells in a 60-mm tissue culture plate to 70–80% confluency. Lyse cells directly in a culture dish by adding 1 ml TRIZOL Reagent to the dish, and passing the cell lysate several times through a pipette. Incubate for 5 min at room temperature. Add 0.2 ml chloroform. Cap sample tubes securely. Shake tubes vigorously by hand for 15 s and incubate them at room temperature for 10 min. 3 Centrifuge the samples at top speed for 30 min at 4°C.
2. Precipitate the RNA from the aqueous phase by mixing with 2.5 volume of cold (kept at –20°C freezer) 100% alcohol to efficiently precipitate miRNA. Incubate samples at room temperature for 10 min and centrifuge at top speed for 10 min at 2°C.
3. When wash the RNA pellet with 75% ethanol, keep at room temperature for over 30 min. Mix the sample by vortexing and centrifuge at no more than $7500 \times g$ for 5 min at 2–8°C.
4. Briefly air-dry the RNA pellet for 5–10 min at room temperature. Dissolve RNA in RNase-free water.

3.3. Enrichment of microRNAs

1. Use 50–100 μg total RNA at concentration of 1 μg/ul in RNA-free water. The RNAs were preheated at 80°C for 3 min, cooled on ice for 2 min (*see* **Note 2**).
2. Apply to top of an YM-100 column. Do not touch the pipette tip on to the membrane and filter through Microcon YM-100 concentrators to obtain enriched microRNAs.
3. Spin at $500 \times g$ at 4°C for 30 min.
4. Save the effluent containing the small RNA in the collection tube. Use a spectrophotometer to assay the material in elution buffer (10 mM Tris, pH 7.5) to obtain an accurate concentration. Reading in water will give inaccurate results.
5. The sample can be frozen at –20°C for several days. Ideally, proceed immediately to the miRNA labeling step (*see* **Note 3**).

3.4. RNA Labeling

1. Use as much as possible of the enriched miRNA derived from 50 to 100 μg total RNA. At room temperature, add the tailing reaction reagents to each miRNA sample in the order shown below and mix well by pipetting a few times. For experiments that include more than samples, it is a good idea to prepare a master mix.

2. Mix the following: 10 μl 5× poly (A) polymerase reaction buffer.
   33 μl RNA (5–10 ug total RNA or as much as possible enriched smwRNA).
   5 μl ^{32}P-α-ATP.
   1 μl poly (A) polymerase.
   1 μl RNasin.

3. Divide the master mix into two 8.5 μl aliquots. Adding 16.5 μl RNAs that contain miRNA to bring the final volume to 25 μl for each sample. Incubate at 37°C for about 2–3 h (*see* **Note 4**).

4. Post-Tailing miRNA Clean-Up is not necessary, but if want, can be done by G25 column filtration (cat. no. PD10, Amersham Pharmacia Biotech).

3.5. Hybridization

1. Prehybridized in 3 ml MicroHyb Hybridization Buffer (Invitrogen) in a hybridization tube at 37°C for at least 30 min, then followed by an overnight hybridization in new hybridization buffer containing RNA probe.

2. Following hybridization, wash membranes twice with 20 ml of 2× SSC/0.5% SDS at 37°C. The second wash was performed twice in 20 ml 1× SSC/0.5% SDS at 37°C. Each for 10 min.

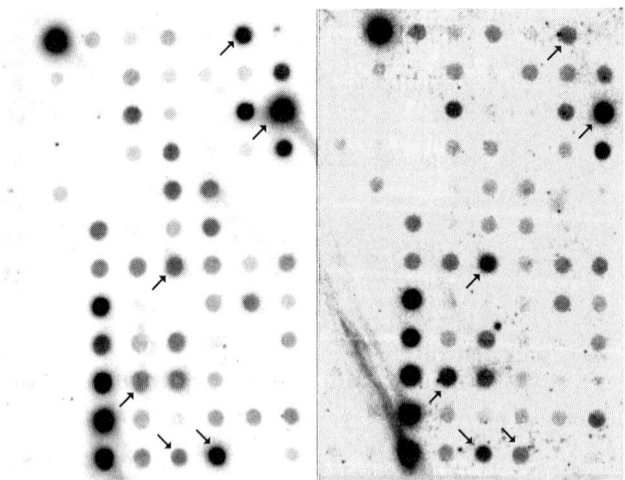

Fig. 1. MicroRNA expression arrays hybridized with RNA from human primary tumor (right panel) or normal tissues (left panel). Arrows show some miRNAs that have significantly different levels in the tumor and normal tissues. A probe complementary to methonine tRNA (tRNA^{Met}) was used as a control for equal hybridization.

3. Expose membranes to a phosphor storage screen, scanned using a Phosphor Imager (Typhoon™) scanner, and quantify hybridization signals using Image Quant software (Molecular Dynamics Inc., Sunnyvale, CA, USA). After exposure, strip the membranes with 0.2% SDS at 72°C, test again by exposure to phosphorimager screen, and rehybridize up to four times.

4. To detect human miRNA differential expression in primary tumor and normal tissues, use these methods, and the results show that there are several miRNA dramatically downregulated or upregulated in tumor tissues (*see* **Fig. 1**).

4. Notes

1. The short nature of microRNAs has presented unique obstacles for miRNA study. Typically, RNA isolation methods that use RNA-binding glass-fiber filters do not quantitatively recover RNA species smaller than 200 nucleotides. Avoid using such methods to extract RNA.
2. It is essential to be preheated at 80°C for 3 min, cooled on ice for 2 min prior to centrifuge. Otherwise, the yield would be very low. Typically, 50 μg total RNA will yield 200 ng smwRNAs.
3. miRNAs are short and thus unstable; labeling should be done immediately after isolation.
4. To label miRNA at high radioactive specificity, we prolong the incubation time to 24 h and got over 1 million cpm/μl.

References

1. Lagos-Quintana, M., Rauhut, R., Lendeckel, W. and Tuschl, T. (2001) Identification of novel genes coding for small expressed RNAs. *Science*, **294**, 853–858.
2. Bartel, D.P. (2004) MicroRNAs: genomics, biogenesis, mechanism, and function. *Cell*, **116**, 281–297.
3. Berezikov, E., Guryev, V., van de Belt, J., Wienholds, E., Plasterk, R.H. and Cuppen, E. (2005) Phylogenetic shadowing and computational identification of human microRNA genes. *Cell*, **120**, 21–24.
4. Lim, L.P., Lau, N.C., Weinstein, E.G., Abdelhakim, A., Yekta, S., Rhoades, M.W., Burge, C.B. and Bartel, D.P. (2003) The microRNAs of Caenorhabditis elegans. *Genes Dev*, **17**, 991–1008.
5. Cullen, B.R. (2004) Transcription and processing of human microRNA precursors. *Mol Cell*, **16**, 861–865.
6. Lee, Y., Ahn, C., Han, J., Choi, H., Kim, J., Yim, J., Lee, J., Provost, P., Radmark, O., Kim, S. et al. (2003) The nuclear RNase III Drosha initiates microRNA processing. *Nature*, **425**, 415–419.
7. Gregory, R.I., Chendrimada, T.P., Cooch, N. and Shiekhattar, R. (2005) Human RISC couples microRNA biogenesis and posttranscriptional gene silencing. *Cell*, **123**, 631–640.

8. He, L. and Hannon, G.J. (2004) MicroRNAs: small RNAs with a big role in gene regulation. *Nat Rev Genet*, **5**, 522–531.

9. Lewis, B.P., Burge, C.B. and Bartel, D.P. (2005) Conserved seed pairing, often flanked by adenosines, indicates that thousands of human genes are microRNA targets. *Cell*, **120**, 15–20.

10. Alvarez-Garcia, I. and Miska, E.A. (2005) MicroRNA functions in animal development and human disease. *Development*, **132**, 4653–4662.

11. Hall, P.A. and Russell, S.H. (2005) New perspectives on neoplasia and the RNA world. *Hematol Oncol*, **23**, 49–53.

12. Schutz, S. and Sarnow, P. (2006) Interaction of viruses with the mammalian RNA interference pathway. *Virology*, **344**, 151–157.

13. Shingara, J., Keiger, K., Shelton, J., Laosinchai-Wolf, W., Powers, P., Conrad, R., Brown, D. and Labourier, E. (2005) An optimized isolation and labeling platform for accurate microRNA expression profiling. *RNA*, **11**, 1461–1470.

14. Nelson, P.T., Baldwin, D.A., Scearce, L.M., Oberholtzer, J.C., Tobias, J.W. and Mourelatos, Z. (2004) Microarray-based, high-throughput gene expression profiling of microRNAs. *Nat Methods*, **1**, 155–161.

15. Liu, C.G., Calin, G.A., Meloon, B., Gamliel, N., Sevignani, C., Ferracin, M., Dumitru, C.D., Shimizu, M., Zupo, S., Dono, M. et al. (2004) An oligonucleotide microchip for genome-wide microRNA profiling in human and mouse tissues. *Proc Natl Acad Sci USA*, **101**, 9740–9744.

16. Sioud, M. and Rosok, O. (2004) Profiling microRNA expression using sensitive cDNA probes and filter arrays. *Biotechniques*, **37**, 574–576, 578–580.

17. Krichevsky, A.M., King, K.S., Donahue, C.P., Khrapko, K. and Kosik, K.S. (2003) A microRNA array reveals extensive regulation of microRNAs during brain development. *RNA*, **9**, 1274–1281.

15

Apoptotic Caspase Activation and Activity

Jean-Bernard Denault and Guy S. Salvesen

Summary

Caspases are central to the execution of apoptosis. Their proteolytic activity is responsible for the demise of cells in many physiological and pathological states. Great advances in understanding caspases have been made using recombinant caspase expression and enzymatic characterization. Assays to measure caspase activity in apoptotic cell extracts and the development of a reconstituted cell-free assay were also critical in establishing the hierarchy in the caspase activation cascade and comprehend how caspase-9 is activated by the apoptosome. More recently, new tools such as activity-based probes allowed us to detect caspase activation in their working environment providing readout of the system with minimal interference. This chapter describes some of the methods used by our group to study the activation mechanisms of caspases and their activity.

Key Words: Activation; activity-based probe; apoptosis; caspase; kinetics.

1. Introduction

Apoptosis is an inflammation-free mechanism by which doomed cells are removed from their environment. It is characterized by a series of hallmarks including nuclear condensation and DNA fragmentation, mitochondrial depolarization, phosphatidyl serine exposure on the plasma membrane, membrane blebbing, and cell packaging into apoptotic bodies. All these phenotypes are caused or regulated by caspases, a family of cysteine proteases, operating by limited proteolysis of a set of cellular proteins (*1,2*). Less than 500 proteins (death substrates) are cleaved by caspases during apoptosis (*2*), but few so far, apart from caspases, have been demonstrated to have essential roles in the

From: *Methods in Molecular Biology, vol. 414: Apoptosis and Cancer*
Edited by: G. Mor and A. B. Alvero © Humana Press Inc., Totowa, NJ

apoptotic process [i.e., inhibitor of caspase-activated DNAse (ICAD) *(3–7)*, poly(ADP-ribose) polymerase (PARP) *(8)*, and Bid *(9,10)*]. Caspases are also important players in specific non-apoptotic roles during inflammatory response *(11–14)*. Other roles of caspases in cell cycle and receptor internalization have been proposed but are less well understood *(15)*.

1.1. Caspase Activation

Caspases are synthesized in the cytosol of all cells as inactive zymogens. A caspase catalytic domain is composed of a large and small subunit linked by a flexible region highly susceptible to proteolysis *(1,16,17)*. Whereas executioner caspases 3, 6, and 7 are dimeric in the cytoplasm, initiator caspases 2, 8, 9, and 10 are monomeric in their inactive conformation *(18,19)*. Simple proteolysis of the executioner caspases is sufficient to gain maximal activity. On the contrary, cleavage of an initiator caspase does not activate it *(18,20)* although initiator caspases are cleaved during apoptosis *(16,21)*. This exemplifies a different mechanism of activation for each class of caspase. In vitro studies of initiator caspases assayed in the presence of kosmotropic salts (such as ammonium or sodium citrate) showed enhanced activity several fold over assay conditions of low salt, whereas very little increase is observed for executioner caspases *(18)*. Furthermore, such salts are able to dimerize caspase-9 *(18,22)*. It is considered that such salts mimic the natural dimerization of initiator caspases that is postulated to occur within activation platforms found in cells *(18,22)*.

Two main cellular pathways trigger apoptosis. The first pathway, termed intrinsic, originates from within the cell to integrate lethal stresses derived from genotoxicity, cell cycle deregulation, cellular traumas, developmental cues, protein misfolding, and various chemical agents. Most, if not all, culminate at the apoptosome, a heptameric macro-molecular complex of AAA+ ATPase homolog Apaf-1 that assembles after the release of cytochrome c from the mitochondria. This complex provides a platform for the activation of caspase-9 *(23–26)*. The second pathway, termed extrinsic, involves the assembly of a cell membrane receptor–adaptor–caspase complex named death-inducing signaling complex (DISC) *(27)*. This assemblage, like the apoptosome, transforms a molecular signal into proteolytic activity, in this case by activating caspase-8 and/or caspase-10. Among the ligands able to activate the DISC are tumor necrosis factor (TNF)-related apoptosis inducing ligand (TRAIL), FasL, and TNFα, which through their cognate receptors and adaptors link the exterior to the cytosol of a cell.

Once active, initiator caspases cleave the linker separating the large and small subunits of the catalytic domain of dimeric executioner caspases and activate them. Using in vitro activation assay, we demonstrated that caspases 8, 9, and 10 strongly prefer Asp 198 of caspase-7 to other potential cleavage/activation sites, the same preference seen for caspase-7 auto-activation in *Escherichia coli* overexpression *(28)*. The precise site within the linker is not of paramount importance for the activation in vitro because granzyme B (GrB), a serine protease delivered by cytotoxic lymphocytes, or exogenous proteases such as subtilisin and cathepsin do activate caspase-7 at different sites *(17)*. In vivo, the cleaved site has some importance because of the newly revealed N-terminus that may be useful for further regulation through degradation *(29)* or inhibition *(30,31)*.

1.2. Caspase Activity Measurement

Caspase activity measurement is a widely used technique in analyzing apoptosis. It should be part of any thorough characterization of cell death process because not all hallmarks described above occur in a given system. Indeed, caspase activation has been proposed to be the central conserved mechanism of apoptosis *(32)*. Because caspases are proteases, it is relatively simple to measure caspase activity using synthetic reporter substrates. Diagnosis of caspase activity is aided by their stringent specificity for cleaving after Asp (P1) (*see* **Note 1**) in proteins and peptide chains. However, the relatively loose extended specificity of caspases makes it difficult to directly associate a cleavage event to a particular caspase. The seminal work of Thornberry and colleagues *(33)* on the specificity of caspases demonstrated that preferences exist for specific sequences, with the caveat that there exist no absolute or defining specificity (*see* **Note 2**). This implies that a given substrate could be the target of many caspases and that some caspases may substitute for others depending on the situation. The same cautions apply to peptidic inhibitors with the added caveat that the irreversible nature of most of those inhibitors tilts the equilibrium toward full inhibition also of other caspases rather than the one originally targeted. Kinetic analysis of small peptidic substrates revealed that caspase-3 is the most proficient caspase with a low K_M and high k_{cat} for many non-optimal substrates *(33,34)*. This means that caspase-3 will overpower other caspases in a mixture such as a cell extract in most settings, even when non-optimal substrates are used such as caspase-8 or caspase-9 "specific" substrates. The bottom line is that no truly specific synthetic substrates or inhibitors are commercially available and that strict caution must be exercised in ascribing observed activity to distinct caspases.

Three main approaches are used to detect caspase activity: detection of cleaved caspase and/or reduced zymogen forms by immunoblotting, enzymatic assays using small peptidic substrates, and activity-based probes (ABPs). In many instances, cleavage of a given caspase constitutes a sign of activation or, at least, a sign that a pathway involving that particular caspase was triggered. It does not demonstrate that a caspase is active, especially for an apical caspase. Moreover, cleaved caspase-3 or caspase-7 could be bound to X-linked inhibitor of apoptosis protein (XIAP), an endogenous caspase inhibitor, and block their activity *(35,36)*; the same could be applied to caspase-9. Thus, the caspases may have been activated, but activity is still held in check by the endogenous inhibitor. This problematic issue is exemplified by the ability of small molecules targeting the endogenous caspase inhibitors to kill tumor cells *(37,38)*.

A more recent development in detecting and measuring caspase activity and activation is the development of ABPs that directly target the substrate-binding site of active caspases. These probes are small peptidic irreversible inhibitors with an affinity purification/detection tag such as biotin or a fluorescent moiety *(39)*. One of the first applications of ABP to study caspases was the use of biotinyl-hexanoic acid-Asp-Glu-Val-Asp-CHO [biotinyl-hexanoic acid-(DEVD)-CHO] for the purification and identification of caspase-3 from THP-1 cells *(40)*. Since then several groups have used and ameliorated them to incorporate irreversible and more specific reactive groups *(39)*. The advantages of using ABPs over immunoblotting and enzymatic assay are several fold. ABPs react primarily with active proteases, which is not the case for antibodies, and, if the APB is cell permeable, will only target non-inhibitor-bound enzyme within their cellular environment. In addition, if ABPs are coupled to immunoprecipitation or immunoblotting, the identity of the labeled enzyme can be revealed along with its molecular form based on size *(28,39,41)*.

In this chapter, we describe a range of techniques upon which several of the studies mentioned above were based. Because we have been asked to focus on our laboratory methodology, readers will find that many of the associated references are from our laboratory.

2. Materials

2.1. Equipment

1. Cell culture 96-well transparent, flat-bottom microplate (Corning Inc. Life science, Lowell, MA, USA).
2. Assay plate 96-well opaque (white), flat-bottom microplate (Corning).
3. Cell culture 96-well transparent, U-bottom microplate (BD Biosciences, Hampton NH, USA).
4. 27-gauge needles with 1-cc syringes.
5. 150-mm tissue culture plates (BD Bioscience Franklin Lakes, NJ, USA).

6. Cell lifter (Fisher Scientific Pittsburg PA, USA).
7. 50-ml disposable conical centrifuge tubes (Corning).
8. 0.22-μm filter units for 500 ml volume.
9. 0.22-μm filter units for 50 ml volume.

2.2. Reagents

1. Purified recombinant caspases (*see* **Note 3**).
2. Caspase dilution buffer: 50 mM Tris, 100 mM NaCl, pH 7.4.
3. High salt caspase buffer (prepared at 1.2 × concentration) 1 ×: 50 mM 4-(2-hydroxyethyl)piperazine-1-ethanesulfonic acid (HEPES), 1 M sodium citrate, 50 mM NaCl, 0.01% w/v 3-[(3-cholamidopropyl)dimethylammonio]-1-propanesulfonate (CHAPS), pH 7.4 (NaOH), and 10 mM dithiothreitol (DTT) (freshly added). Notice this buffer uses a very high citrate concentration.
4. Caspase buffer (prepared as 2× concentration): 20 mM 1,4-piperazinediethanesulfonic acid (PIPES), 200 mM NaCl, 20% w/v sucrose, 0.2% w/v CHAPS, 2 mM ethylenediaminetetraacetic acid (EDTA), pH 7.2 (NaOH), and 20 mM DTT (freshly added).
5. 1 M DTT in double-distilled water (ddH$_2$O); store at –20°C in small aliquot.
6. 20 mM Acetyl-Asp-Glu-Val-Asp-paranitroanilide (AcDEVD-pNA) substrate or other pNA chromogenic substrates (Bachem Bioscience, King of Prussia, PA, USA) in dimethylsulfoxide (DMSO).
7. 10 mM AcDEVD-7-amino-4-trifluoromethyl coumarin (AcDEVD-Afc) substrate or other Afc fluorogenic substrates (MP Bioscience, Solon, OH, USA) in DMSO.
8. Sterile ddH$_2$O.
9. 10 mM carbobenzoxy-Val-Ala-Asp-fluoromethyl ketone (zVAD-fmk) inhibitor (MP Bioscience) in DMSO.
10. Hypotonic buffer (prepared at 5× concentration) 1 ×: 20 mM PIPES, 20 mM KCl, 5 mM EDTA, 2 mM MgCl$_2$, pH 7.4 (NaOH), and 2 mM DTT (freshly added).
11. Complete Dulbecco's modified Eagle medium (DMEM Mediatech Inc. Herndon, VA, USA) supplemented with 10% v/v heat-inactivated fetal bovine serum (FBS), 2 mM L-glutamine, 100 U/ml penicillin, and 100 μg/ml streptomycin (Invitrogen, Carlsbad, CA, USA).
12. Complete RPMI 1640 media (Mediatech Inc.) supplemented in the same way as for DMEM above (**item 11**).
13. 100 μM cytochrome c (from horse heart, Sigma-Aldrich, St-Louis MO, USA) in ddH$_2$O; store at –20°C.
14. 100 mM 2′-deoxyadenosine 5′-triphosphate in ddH$_2$O (dATP, Roche, Indianapolis, IN, USA).
15. mRIPA–EDTA buffer: 50 mM Tris, 100 mM NaCl, 1% v/v NP-40, 0.5% w/v deoxycholic acid, 0.1% w/v sodium dodecyl sulfate (SDS), and 1 mM EDTA (pH 7.4).
16. 50% w/v sucrose in ddH$_2$O; filter-sterilize and keep at 4°C.

17. 40% w/v acrylamide:bis-acrylamide (37.5:1) solution (Fisher Scientific, Pittsburg, PA, USA).

18. 40% w/v acrylamide:bis-acrylamide (19:1) solution (Sigma-Aldrich).

19. 10% w/v ammonium persulfate (APS, Fisher Scientific) in ddH$_2$O; keep at 4°C.

20. N,N,N′,N′-tetramethylethylenediamine (TEMED, BioRad, Hercules CA, USA).

21. 5× Lower gel buffer (LGB): 556 mM 2-amino-2-methyl-1,3-propandiol (ammediol, Sigma-Aldrich) and 224 mM HCl (no need to adjust pH); filter-sterilize and keep at 4°C.

22. 4× Upper gel buffer (UGB): 334 mM ammediol and 239 mM HCl (no need to adjust pH); filter-sterilize and keep at 4°C.

23. Low acrylamide mix (8%, 50 ml): 20 ml ddH^{2}O, 10 ml 50% sucrose, 10 ml 5× LGB, 10 ml 40% acrylamide (37.5:1), 75 μl 10% APS, and 30 μl TEMED.

24. High acrylamide mix (18%, 50 ml): 7.5 ml ddH$_2$O, 10 ml 50% sucrose, 10 ml 5× LGB, 22.5 ml 40% acrylamide (37.5:1), 75 μl 10% APS, and 30 μl TEMED.

25. Ethanol–LGB solution: 25% v/v ethanol in 1× LGB.

26. Stacking gel (4 ml): 1.68 ml ddH$_2$O, 1 ml 50% sucrose, 1 ml 4× UGB, 0.32 ml 40% acrylamide (19:1), 40 μl 10% APS, and 4 μl TEMED.

27. 5× Upper reservoir buffer (URB): 205 mM ammediol, 200 mM glycine (BioRad), and 0.5% w/v SDS (BioRad).

28. 5× Lower reservoir buffer (LRB): 313 mM ammediol and 240 mM HCl.

29. 3× Gel loading buffer: 50 ml 4× UGB, 20 ml glycerol, 7 g SDS, 0.1% w/v bromophenol blue, and 60 mM DTT (freshly added).

30. SDS–polyacrylamide gel electrophoresis (SDS–PAGE) broad-range molecular weight markers (BioRad).

31. Transfer buffer: 10 mM 3-[cyclohexylamino]-1-propanesulfonic acid (CAPS) and 10% v/v methanol [high-performance liquid chromatography (HPLC) grade].

32. Ponceau S solution: 0.25% w/v Ponceau S (Sigma-Aldrich) in 1% v/v acetic acid.

33. PBS (prepared at 10× concentration) 1×: 137 mM NaCl, 2.7 mM KCl, 10.2 mM Na$_2$HPO$_4$, and 1.76 mM KH$_2$PO$_4$, pH 7.4 (HCl or NaOH).

34. PBS-T: PBS containing 0.1% v/v Tween-20 (BioRad).

35. PBS-T-milk: PBS containing 5% w/v non-fat dry milk (Carnation Nestlé, Wilkes-Barre, PA, USA) and 0.2% v/v Tween-20.

36. PBS–BSA: PBS containing 3% w/v bovine serum albumin (BSA) Fraction V (Sigma-Aldrich).

37. Anti-mouse and anti-rabbit IgG secondary antibodies (GE Healthcare, Piscataway, NJ, USA).

38. WestPico SuperSignal chemiluminescence detection kit (Pierce Chem. Co., Rockford, IL, USA).

39. PBS-T-iBlock: PBS containing 0.2% w/v iBlock (dissolved at 50–60°C, Tropix) and 0.1% v/v Tween-20 (added once PBS-iBlock has cooled). For convenience, iBlock can be replaced by 3% BSA fraction V.

40. Streptavidin–horse-radish peroxidase (HRP) (powder, Sigma-Aldrich) prepared at 1 mg/ml in PBS containing 50% v/v glycerol; keep at –20°C in small aliquot.

3. Methods

The techniques described below include (i) the measurement of executioner and initiator caspases activity, (ii) titration of recombinant caspases, (iii) measurement of caspase activity in cell extracts, (iv) activation of caspase in cell extracts and in vitro, and (v) labeling of caspases with ABPs in whole cells and extracts, and of recombinant enzymes.

3.1. Activity Measurement of Recombinant Caspases In Vitro

Because caspases are obtained at high concentration during *E. coli* expression, initiator caspase preparations are often a mixture of dimer and monomer. For example, caspase-8 can be obtained as the active dimer at concentrations of >10 μM after purification because the reported K_D is 3–5 μM *(42)* in low salt buffer. In vitro high concentrations of kosmotropic salts were found to force dimerization by lowering K_D by 10-fold to 100-fold, hence activating initiator caspases *(18)*. For executioner caspases, incubation at 37°C to warm the enzyme in the presence of DTT is sufficient to obtain full enzymatic activity. **Figure 1** presents a routine kinetic analysis of caspase-7 and caspase-8 to obtain kinetic parameters in low and high salt conditions. Caspase preparations for these assays were recombinant caspase-7 (dimeric) and monomeric recombinant caspase-8 obtained by anion exchange. High salt has a relatively weak effect on caspase-7 increasing its k_{cat} twofold, whereas caspase-8 is more affected with a 23-fold increase in k_{cat} and fourfold lowered K_M. Hence, the activation (k_{cat}/K_M) of caspase-7 by salt is only twofold, whereas caspase-8 activation is >90-fold. This is similar to the activation reported by Boatright and colleagues *(18)*.

3.1.1. Executioner Caspases

1. Thaw a 1 M DTT and caspase aliquots on ice.
2. Add DTT to 20 mM in 2× caspase buffer (2 ml); keep on ice.
3. Dilute caspase stock using caspase dilution buffer (*see* **Notes 2** and **4**); keep on ice.
4. In an opaque white flat-bottom 96-well plate, dispense 50 μl of 2× caspase buffer. Opaque black plates may also be used, but fluorescence yields will be much lower.
5. Add diluted caspase to each well (*see* **Note 2**).
6. Include a control well with caspase dilution buffer alone.
7. Top up to 80 μl with caspase dilution buffer; mix well by quickly pipetting up and down three to five times with a multi-channel pipette.

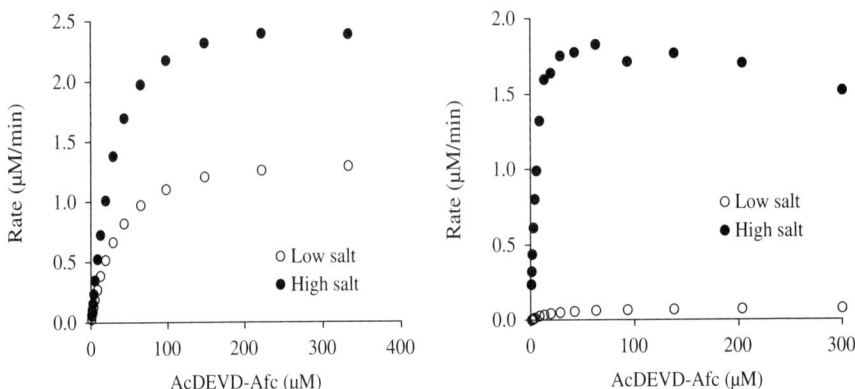

Fig. 1. (**A**) 2.5 nM recombinant caspase-7 or (**B**) 20 nM recombinant caspase-8 was activated in a low salt (open circles, 50 mM HEPES, pH 7.0, 50 mM NaCl and 0.01% CHAPS) or high salt buffer (close circles, buffer with 1 M sodium citrate; buffer with 0.7 M sodium citrate for caspase-8) and was incubated with various concentrations of acetyl-Asp-Glu-Val-Asp-7-amino-4-trifluoromethyl coumarin (AcDEVD-Afc) or AcIETD-Afc, respectively (*see* **Subheadings 3.1.1.** and **3.1.2.**). The initial rates obtained from each sample using continuous measurement were plotted against substrate concentration to obtain a typical Michaelis–Menten substrate saturation curve. Determined K_M and k_{cat} for caspase-7 were 33.8 μM and 9.6 per second (low salt) and 30.1 μM and 18.1 per second (high salt); K_M and k_{cat} parameters for caspase-8 were 19.8 μM and 0.07 per second (low salt) and 5.2 μM and 1.6 per second (high salt).

8. Incubate at 37°C for 15 min. This is best done in the plate reader.
9. Dilute Afc fluorogenic substrate to 2500 μM in dilution buffer (25 μl per sample); warm to 37°C. Other concentrations of substrate (in the range 1–2500 μM) could be used to determine kinetic parameters of substrate hydrolysis (e.g., K_M).
10. Add 20 μl of substrate to each well (final concentration of 100 μM for a routine assay); thoroughly mix.
11. Read immediately at 37°C in kinetic/continuous mode with intervals as short as possible (EX$_\lambda$: 405 nm, EM$_\lambda$: 510 nm) (*see* **Note 5**).

3.1.2. Initiator Caspases

1. Thaw a 1 M DTT and caspase aliquots on ice.
2. Add DTT to 12 mM in 1.2× high salt buffer (see **Subheading 2.2., item 3**) (2 ml); keep at room temperature.
3. Add 100 nM initiator caspase to the prepared 1.2× high salt buffer and dilute buffer to 1× concentration with ddH$_2$O (0.5 ml). The initiator caspase concentration should be 100 nM final (*see* **Note 4**).

4. Incubate at 37°C for 1 h.
5. In an opaque white flat-bottom 96-well plate, dispense the desired amount of activated initiator caspases (*see* **Note 2**) planning for a final volume of 100 μl. Opaque black plates may also be used, but fluorescence yields will be much lower.
6. Include a control well with caspase dilution buffer alone.
7. Top up to 80 μl with 1× assay buffer; mix well by repeatedly pipetting up and down three to five times with a multi-channel pipette.
8. Follow **steps 9–11** above (*see* **Subheading 3.1.1.**).

3.1.3. Caspase Titration

Caspase activities are frequently cited in terms of units in the literature, but we advise against this because each publication has a different definition of "unit". The preferred way to describe enzyme activity is active site concentration, because this provides a more reliable comparison between different laboratories. Therefore, in all assays, the exact concentration of active enzyme should be determined if possible. This is important because caspase preparations vary in purity, quality, and activity. Moreover, it is necessary to obtain accurate k_{cat} values and other kinetic parameters.

Caspase titration is a two-step process. First, the caspase is incubated with the irreversible inhibitor (the titrant) zVAD-fmk at concentrations 0–10 times the provisional caspase concentration in a minimal volume in assay conditions. The provisional caspase concentration is estimated by the Edelhoch relationship, which relates the amino acid composition of a given protein to its absorbance at 280 nm (*43*). After preincubation, the assay is diluted by the addition of an excess of substrate (100–200 μM) and the residual activity is determined as in a standard assay. The concentration of active caspase is obtained by plotting the relative activity rate against the concentration of zVAD-fmk. The intercept of the tangent to the slope with the *x*-axis gives the active enzyme concentration in the assay. The accurate titration is achieved when all the titrant has reacted with the enzyme. This rate of the reaction is facilitated at high enzyme concentration. Chromogenic substrates are thus preferred in this application rather than fluorogenic substrates because they are less sensitive and allow the usage of a higher enzyme concentration. Further details could be found in **ref. 44**. The same protocol can be used for initiator caspases using high salt buffer and with a preincubation period of 1 h to allow complete activation (*see* **step 5** below).

Figure 2 shows the full range of zVAD-fmk concentrations used to determine the concentration of active caspase-3 and the restricted values used to determine the active concentration. The assay was designed so caspase-3 was at 100 nM

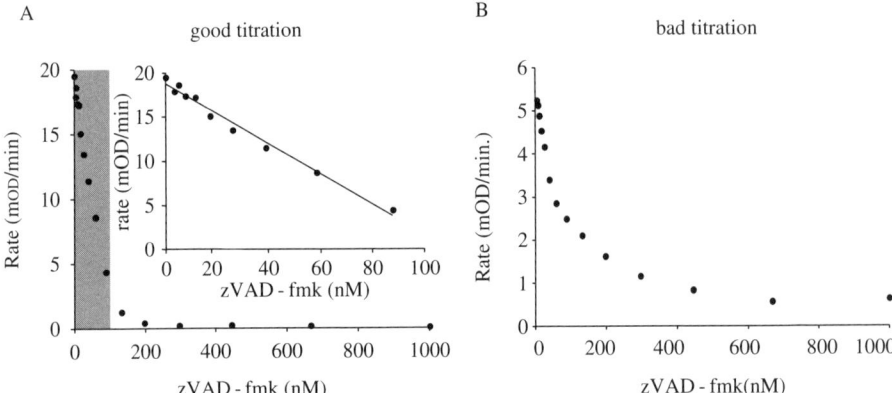

Fig. 2. **(A)** Caspase-3 was titrated using carbobenzoxy-Val-Ala-Asp-fluoromethyl ketone (zVAD-fmk) as titrant and acetyl-Asp-Glu-Val-Asp-paranitroanilide (AcDEVD-pNA) as a substrate to measure remaining activity (*see* **Subheading 3.1.3.**). The titration reaction was done using 100 nM caspase-3 based on protein concentration, and a 10-fold dilution of the inhibited reaction was used to measure the remaining activity. The final value was found to be 109.5 nM in the first step of the assay. The inset shows the data points (shaded area) used to determine the active caspase-3 concentration. **(B)** Example of a bad titration of caspase-10 using zVAD-fmk. No clear linear portion can be used to get the active caspase concentration. In addition, complete inhibition is never attained even at 1 μM titrant.

protein concentration based on the Edelhoch relationship. Because caspase-3 is highly active on AcDEVD-pNA, dilutions of the titration assays were used in the second step of the assay. The final value was found to be 109.5 nM in the first step of the assay. This type of titration allows an accurate determination of the actual active site concentration and corrects the estimated concentration. It relies completely on an accurate value of the titrant (zVAD-fmk) concentration, so care must be taken to be precise when reagent is weighed and dissolved in the stock solvent.

1. Estimate the concentration of caspase in the sample you plan to titrate using the Edelhoch relationship.
2. Thaw a 1M DTT and caspase aliquots on ice.
3. Add DTT to 10 mM in 1× caspase buffer (2 ml).
4. Prepare 200 nM caspase in a microfuge tube in 1× caspase buffer (200 μl).
5. Incubate at 37°C for 15 min.
6. In a clear flat-bottom 96-well plate, pipet 10 μl of a serial dilution of zVAD-fmk prepared in 1× caspase buffer into 15 wells to cover a 1000-fold concentration

range starting at 2 μM as the highest titrant concentration. A two-third serial dilution starting at 2 μM is suitable.

7. Include a sample without inhibitor for the uninhibited control sample.
8. Transfer 10 μl of diluted enzyme to each well; mix well by shaking in the microplate reader or by pipetting up and down three to five times with a multi-channel pipette.
9. Incubate for 30 min at 37°C (15 min for caspase-6; 1 h for caspase-2).
10. Prepare 250 μM AcDEVD-pNA substrate in 1× caspase buffer (1.5 ml); warm to 37°C.
11. With a repeater pipette, rapidly add 80 μl of diluted pNA substrate; mix thoroughly.
12. Read immediately at 37°C in kinetic/continuous mode with intervals as short as possible (EX_λ: 405 nm, EM_λ: 510 nm; read at 405 nm for chromogenic substrate) (*see* **Notes 5** and **6**).
13. Plot the residual activity against zVAD-fmk concentration and determine the linear regression for values at low zVAD-fmk concentrations including as many data points as possible. The intercept of the slope with the *x*-axis gives the concentration of enzyme in the assay in **step 9**.
14. If desired, titration can be repeated with a narrower range of zVAD-fmk concentrations to obtain a more accurate active caspase concentration.

3.2. Activity Measurement in Extracts

Although many investigators rely on cleavage of caspases as readout of their activation, this can be misleading, particularly if one wishes to measure initiator caspases. Therefore, direct measurement of caspase activity by using synthetic reporter substrates is preferred. But there are caveats to the interpretation of caspase activity measurements.

The activity of initiator caspases in cell extracts is minimal because preparation dilutes them, hence favoring inactivation of the initiator caspases through dissociation of the active dimer. Furthermore, the activity of executioner caspases surpasses the one from initiator caspases. This means that reporter substrates will usually measure the activity of executioner caspases. Finally, most extract preparation techniques do not protect the integrity of initiator caspase activation platforms (DISC, apoptosome, PIDDosome, or inflammasome), and so accurate determination of the activity of their target initiator caspases may be underestimated.

Nevertheless, one of these platforms, the apoptosome, can be recapitulated in vitro in hypotonic extract *(45,46)*. This technique, which uses the general caspase substrate AcDEVD-pNa, was used to study the activation of caspase-9 in controlled conditions. For example, Stennicke and colleagues used it to demonstrate that cleavage is not required or sufficient to activate caspase-9

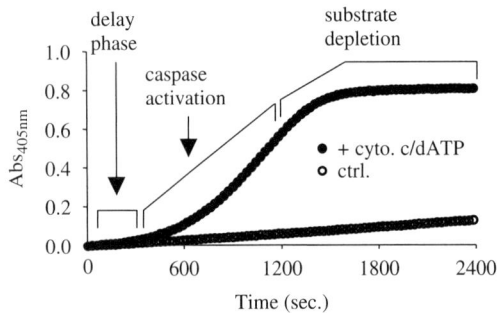

Fig. 3. (A) Hypotonic extract was activated with 1 μM cytochrome c and 1 mM dATP (*see* **Subheading 3.2.1.**); 400 μM acetyl-Asp-Glu-Val-Asp-paranitroanilide (AcDEVD-pNA) was used to monitor caspase-3/caspase-7 activation. The three phases displayed in the assay are identified. Note the exponential aspect of the activation section of the trace. The steepness of the slope is indicative of the strength and quality of the extract. At the end of the assay, the amount of AcDEVD-pNA consumed in the unactivated control is roughly 20% that of the fully depleted sample.

(20). More recently, our group used reconstituted hypotonic extract to demonstrate that the caspase recruitment domain (CARD) of caspase-9 is sufficient for the activation of a caspase-8 hybrid *(22)*. The basic protocol is described below (*see* **Subheading 3.2.1.**). A typical result of hypotonic extract activation is presented in **Fig. 3** and shows several phases in the activation process.

3.2.1. In Hypotonic Extract

3.2.1.1. Cell Culture

1. Grow 293A or 293T cells in complete DMEM and passage them at 1:8 when confluence reaches 90%. 293A, an adherent clone of the original HEK-293 cell line used to propagate adenovirus, is available form Q-Biogene (Irvine CA, USA) whereas 293T are HEK-293 cells expressing the SV40 T-antigen and was developed in Michelle Calos laboratory (Stanford University).
2. Seed cells into 10–20 150-mm tissue culture plates and grow them to a confluency of no more than 70%. Higher confluency gives unacceptable backgrounds of caspase activity.
3. Aspirate media and harvest cells with a cell lifter in 5 ml of ice-cold PBS. Subsequent steps must be carried out on ice.
4. Collect cells by centrifugation at 500 × *g* for 10 min at 4°C in a 50-ml conical tube.
5. Discard supernatant.

3.2.1.2. PREPARATION OF HYPOTONIC EXTRACT

1. Resuspend cells by gently pipetting with a wide bore pipette in 10 ml of ice-cold PBS; harvest as above (*see* **Subheading 3.2.1.1., step 3**).
2. Prepare 1× hypotonic buffer (without DTT); keep on ice.
3. Thaw a 1 M DTT aliquot.
4. Resuspend cells in one cell pellet volume of ice-cold hypotonic buffer (without DTT).
5. Transfer to a 1.5-ml microfuge tube.
6. Collect cells by centrifugation at 2000 × *g* for 1 min at 4°C.
7. Add DTT to 2 mM in 1× hypotonic buffer (2 ml); keep on ice.
8. Resuspend cells in one cell pellet volume of ice-cold hypotonic buffer (with DTT).
9. Incubate on ice for 20 min.
10. Pass cells 20 times through a 27-gauge needle attached to a 1-cc syringe.
11. Clear cell debris by centrifugation at 18,000 × *g* for 30 min at 4°C.
12. Use immediately or aliquot supernatant in 45-μl portions and freeze at –80°C.

3.2.1.3. ACTIVATION AND MEASUREMENT OF ACTIVITY IN HYPOTONIC EXTRACT

1. Thaw 2 aliquots of hypotonic extract on ice.
2. Prepare a 50 mM dATP and 50 μM cytochrome c solution in ddH$_2$O (10 μl).
3. In a clear U-bottom 96-well microplate, dispense 40 μl of hypotonic extract.
4. Include a well as a control that will not be activated.
5. Warm the plate to 37°C for 5 min.
6. Add 2 μl of 20 mM AcDEVD-pNA substrate to each well.
7. Add 1 μl of cytochrome c and dATP mix; mix thoroughly.
8. Read immediately at 37°C in kinetic/continuous mode with intervals as short as possible (absorbance at 405 nm) (*see* **Note 5**).

3.2.2. In Detergent Extract

The above protocol cannot be used if samples already contain active caspases and cannot be used in transfection experiments because of the large amount of cells required to prepare concentrated hypotonic extracts. A means to measure caspase activity in small samples is hence useful. Although many kits are commercially available, they are expensive and can be simple to make from scratch because they consist of a lysis buffer and a fluorogenic or chromogenic caspase substrate.

Cells cultivated in a 60-mm plate provide sufficient material for an enzymatic assay and one to five immunoblots depending on the sensitivity of antibodies. Cells could be treated to undergo apoptosis or transfected with specific cDNA to alter the apoptotic program. For example, Scott and colleagues used such assay to assess the inhibitory effect of various XIAP mutants on the activity of caspases in whole cells treated with TRAIL *(30) (see* **Fig. 4**).

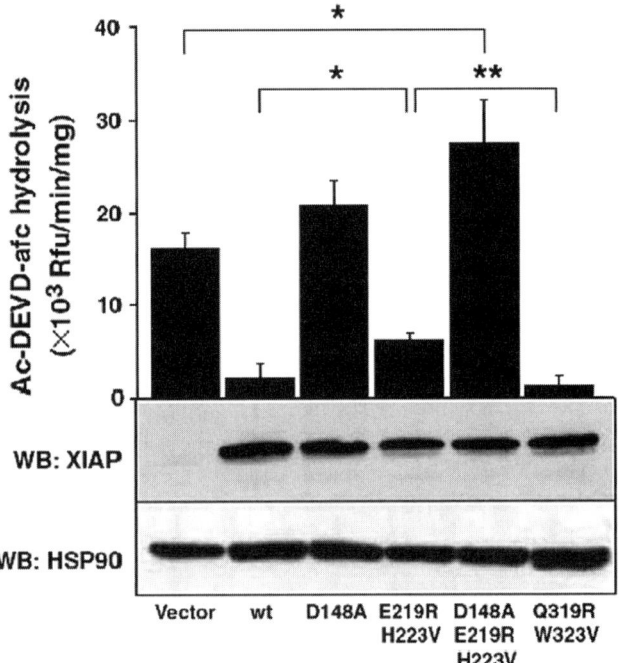

Fig. 4. 293A cells were transfected with 0.5 μg of myc-XIAP plasmids and treated with 100 ng/ml TRAIL for 2 h, and lysates were prepared in mRIPA-EDTA buffer (*see* **Subheading 3.2.2.**). One-tenth of the lysate was added to an assay containing 100 μM acetyl-Asp-Glu-Val-Asp-7-amino-4-trifluoromethyl coumarin (AcDEVD-Afc). Initial rates were normalized for protein concentration. The lysates were balanced for equal protein concentration and analyzed by immunoblotting with monoclonal anti-XIAP or anti-hsp90 antibody as a loading control (*see* **Subheading 3.5.**) *(30)*.

3.2.2.1. Preparation of Detergent Extract

1. Cultivate 293A cells in 60-mm dishes in complete DMEM to a confluence of 60%.
2. Transfect cells with 0.5 μg of plasmid DNA of interest using Fugene 6 (or other transfection reagents) as described by the suppliers.
3. Twenty-four hours post transfection, treat cells with 100 ng/ml TRAIL for 2 h (or other stimuli and treatment conditions).
6. Aspirate media and harvest cells in 1 ml of ice-cold PBS by pipetting.
7. Collect cells by centrifugation at 500 × *g* for 10 min at 4°C in a microfuge tube.
8. Discard supernatant.
9. Resuspend cell pellet in 100 μl of mRIPA–EDTA lysis buffer with a pipette.

10. Incubate on ice for 10 min (*see* **Note 7**).
11. Clear cell debris by centrifugation at 18,000 × *g* for 20 min at 4°C.

3.2.2.2. MEASUREMENT OF ACTIVITY IN DETERGENT EXTRACT

1. Thaw a 1 M DTT and caspase aliquots on ice.
2. Add DTT to 20 mM in 2× caspase buffer (1 ml).
3. Dilute AcDEVD-Afc substrate to 1 mM in ddH$_2$O (12 µl per sample).
4. In an opaque flat-bottom 96-well microplate, transfer 50 µl of 2× caspase buffer for each assay to be performed; warm to 37°C for 5 min.
5. Transfer 40 µl of cleared lysate (*see* **Subheading 3.2.2.1., step 7**) to a well containing buffer; warm to 37°C for 5 min.
6. Add 10 µl of diluted substrate to each well.
7. Read immediately at 37°C in kinetic/continuous mode with intervals as short as possible (EX$_\lambda$: 405 nm, EM$_\lambda$: 510 nm) (*see* **Note 5**).

3.3. Activation of Recombinant Caspases In Vitro

The mechanism by which initiator and executioner caspases are activated differs (*see* **Subheading 1.**). Because initiator caspases can be activated in vitro by incubating them in high salt conditions, the assay described in **Subheading 3.1.1** is an activation assay. Activation of hypotonic extracts (*see* **Subheading 3.2.1.**) is also a way to activate caspase-9 in vitro, but caspase-9 activity cannot be accurately measured because caspase-3 and caspase-7 are present. For executioner caspases, which can be obtained as zymogens by short expression times (uncleaved, inactive) in *E. coli*, proteolysis of the linker region with an initiator caspase (*see* **Subheading 3.3.1.1.**) or GrB (*see* **Subheading 3.3.1.2.**) will result in activation of the executioner caspases.

3.3.1. Executioner Caspases

In vitro activation of executioner caspases is achieved when a protease cleaves the linker region of the caspase zymogen. Activation is detected using a good peptidic substrate for the executioner caspase. For the activation by an initiator caspase, high salt caspase buffer is used to allow initiator caspases to reach full activity before adding to the executioner caspase zymogen.

The trick to this protocol is determining the optimal concentrations of the zymogen executioner caspase to be activated and the amount of activator. This is also complicated by the fact that many activators, especially caspases 8 and 10, have relatively good activity on the reporter substrate used to detect executioner caspase activation at the concentration necessary to efficiently activate executioner caspases.

In the example described in **Fig. 5A,** a series of concentrations of pro-caspase-7 are activated by a fixed quantity of caspase-9. The chromogenic substrate AcDEVD-pNA traces the activated caspase-7. Chromogenic substrate is preferred to the more sensitive fluorogenic substrates because it allows the use of higher concentrations of activator and executioner caspases. We used this type of activation assays to determine the ability of caspases 2, 8, 9, and 10 to activate pro-caspase-7 *(47)*. It has also been used with other activators such as GrB or by exogenous proteases such as cathepsins and subtilisin for pro-caspase-3 and pro-caspase-7 *(17,48)*. The second protocol allows conversion of pro-caspase-7 into fully processed caspase using GrB as an activator. This is especially useful if one wants to convert an inactive, single-chain caspase form into a processed two-chain form (*see* **Fig. 5B**).

3.3.1.1. ACTIVATION OF EXECUTIONER CASPASES BY INITIATOR CASPASES

1. Thaw a 1M DTT and caspase aliquots on ice.
2. Add DTT to 10 mM in 1× high salt caspase buffer (2 ml).

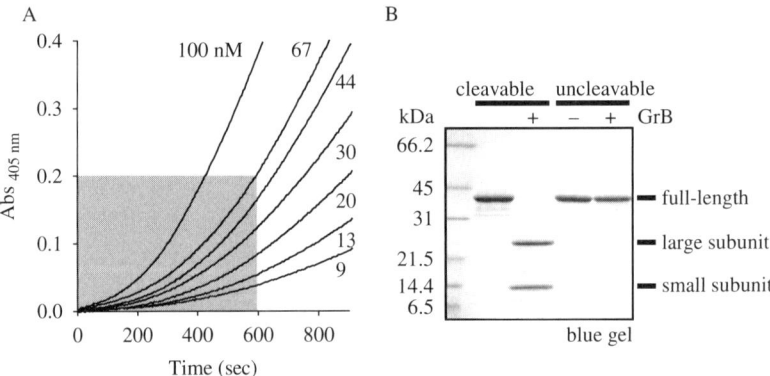

Fig. 5. **(A)** Various concentrations of pro-caspase-7 (9–100 nM) were incubated with 50 nM caspase-9 in high salt caspase buffer, and acetyl-Asp-Glu-Val-Asp-paranitroanilide (AcDEVD-pNA) was use to follow the activation of caspase-7 (*see* **Subheading 3.3.1.1.**). Note that the progress curves have an exponential aspect in the early phase, and this is the phase that must be used to calculate activation rates (shaded area). The sharper the rate of AcDEVD-pNA hydrolysis increases, the stronger the activation process. **(B)** The catalytic mutant of caspase-7 with an intact linker region (cleavable) or the canonical activation site mutant (uncleavable) was incubated in buffer alone (–) or with granzyme B (GrB) (+) and analyzed by SDS–PAGE (*see* **Subheading 3.3.1.2.**) *(47)*.

3. Prepare 100 nM caspase-9 in high salt caspase buffer and incubate for 1 h at 37°C (0.5 ml).
4. Prepare 4 mM AcDEVD-pNA in high salt caspase buffer (12 μl per assay) and warm to 37°C.
5. Ten minutes before caspase-9 activation is over, set a serial dilution of pro-caspase-7 in a column of a clear flat-bottom 96-well plate (wells B through H) starting with 250 nM and down to 25 nM in 40 μl of high salt caspase buffer.
6. Add 40 μl of 1× high salt caspase buffer to a well (A) for the activator alone sample.
7. Incubate the plate at 37°C for 5 min.
8. Using a repeater pipette, quickly add 10 μl of substrate to each well.
9. Using a repeater pipette, quickly add 50 μl of activated caspase-8 to each well and mix rapidly.
10. Read immediately at 37°C in kinetic/continuous mode with intervals as short as possible (absorbance at 405 nm) (*see* **Note 5**).

3.3.1.2. ACTIVATION OF EXECUTIONER CASPASES BY GRANZYME B

1. Thaw an aliquot of GrB and a pro-caspase (*see* **Note 3**) aliquot on ice.
2. Prepare 1 μg of pro-caspase in caspase dilution buffer.
3. Add 10 ng of GrB to the reaction (this is a concentration ratio of 1:100 based on protein amount).
4. Incubate for 30 min at 37°C.
5. If the pro-caspase is expected to gain activity upon cleavage, perform an assay as described in **Subheading 3.1.2**.
6. Analyze the processing efficacy on SDS–PAGE using at least 200 ng of protein (*see* **Subheading 3.5.**, the gel is stained using GelCode blue after **Subheading 3.5.1.3**). Smaller amounts (10 ng) can be analyzed by immunoblotting using caspase-specific antibody (*see* **Subheading 3.5.**).

3.4. Labeling of Caspases Using Activity-Based Probes

ABPs consist of three functionalities: (i) an electrophilic group, such as a fmk or acyloxymethyl ketone (aomk) that irreversibly labels the catalytic site, (ii) a peptidyl sequence that targets a specific group of proteases, and (iii) a moiety for detections—such as biotin or a fluorophore. For caspases, probes with an Asp adjacent to the electrophile are used *(39)*. As a rule of thumb, 1–10 μM of the probe (bVAD-fmk or bEVD-aomk) is sufficient to label any caspases in vitro given enough time because of the irreversible nature of these ABPs. **Figure 6** presents the labeling of caspase-3 and caspase-7 in extract of cells treated with anti-Fas antibodies and etoposide. The experiment uses bEVD-aomk (*see* **Note 8**) to label an apoptotic extract, and labeled proteins are isolated with streptavidin beads. Individual caspase is detected using a specific

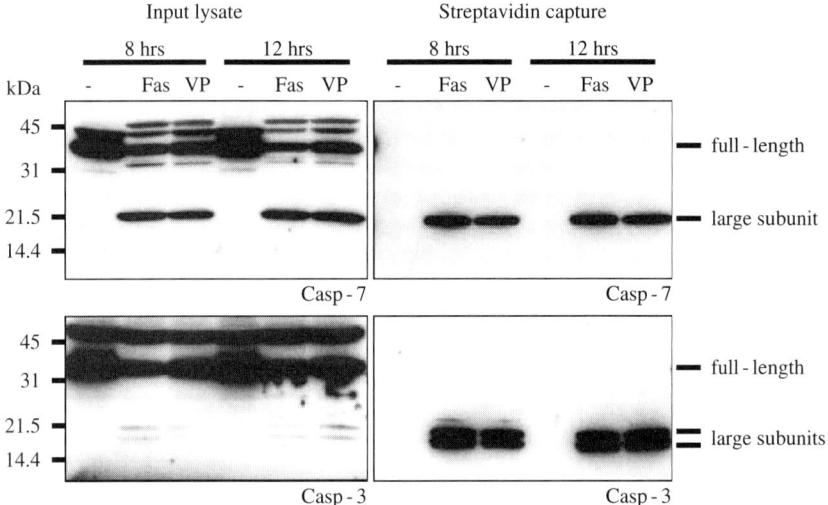

Fig. 6. Jurkat cells were treated with anti-Fas antibodies (50 ng/ml) or VP-16 (50 μM) for 8 or 12 h (*see* **Subheading 3.4.1.2.**). Hypotonic extracts were prepared in the presence of protease inhibitors and 1 μM bEVD-acyloxymethyl ketone (biotinyl-hexanoic acid-GluVal-Asp (bEVD)-aomk) activity-based probe (ABP). Input lysate (5%) and pull-down proteins (25%) were analyzed by immunoblotting using caspase-3 and caspase-7 antibodies.

antibody, and labeled proteins can be detected using HRP-coupled streptavidin (*see* **Subheading 3.5.**). In **Fig. 6**, active caspase-7 is a 20-kDa form, whereas caspase-3 has a 20 and 17 kDa corresponding to the expected active form of both executioner caspases. This type of experiment was used to demonstrate that both the DISC and the apoptosome are very effective at cleaving the catalytic units of both caspase-3 and caspase-7 dimers in a recently published article *(47)*.

3.4.1. Labeling Apoptotic Extracts

3.4.1.1. Cell Culture

1. Grow Jurkat cells in complete RPMI 1640 media and maintain at $0.1–2 \times 10^6$ cells/ml.
2. For each sample, grow 10^7 cells to a density of 10^6 cells/ml.
3. Collect cells by centrifugation at $200 \times g$ for 10 min at room temperature; discard supernatant.
4. Gently resuspend cell pellet in complete culture media at 2×10^6 cells/ml.

5. Treat cells with 50 ng/ml anti-Fas (CH-11) antibody or 50 μM etoposide or leave untreated; replace into cell culture incubator for 8–12 h.
6. Harvest cells as above.
7. Wash cells with 10 ml of cold PBS; harvest cells as above.
8. Transfer cells to a microfuge tube using 1 ml of cold PBS.
9. Drain cell pellet with a 27-gauge needle attached to a vacuum line.

3.4.1.2. PREPARATION OF EXTRACT AND LABELING

1. Thaw a 1 M DTT aliquot.
2. Add DTT to 2 mM in 1× hypotonic buffer (1 ml).
3. Resuspend cells in 0.1 ml of hypotonic buffer containing 1 μM bEVD-aomk.
4. Incubate at 37°C for 45 min.
5. Pass lysate through a 27-gauge needle 10 times.
6. Incubate at 37°C for 15 min.
7. Repeat **step 5** above.
8. Clear lysate by centrifugation at $18,000 \times g$ for 30 min at 4°C.
9. Transfer supernatant to a fresh tube.
10. Reserve 30 μl of sample for input control immunoblots.

3.4.1.3. CAPTURE USING STREPTAVIDIN–HRP

1. Add 0.33 ml of mRIPA buffer containing protease inhibitors (*see* **Note 9**) and caspase inhibitors (100 μM AcDEVD-CHO and 100 μM zVAD-fmk).
2. Add 20 μl of mRIPA-washed streptavidin beads in 0.1 ml of mRIPA buffer to each sample.
3. Agitate gently overnight at 4°C.
4. Recover beads by centrifugation at $800 \times g$ for 1 min at 4°C.
5. Wash beads three times with 1 ml cold mRIPA buffer; recover beads as above.
6. Wash beads twice with cold PBS; recover beads as above.
7. Drain beads with a 27-gauge needle attached to a 1-cc syringe.
8. Boil beads in 50 μl of SDS–PAGE loading buffer containing 1 mM d-biotin from 10 min. Adding 1 mM d-biotin can improve recovery at this step.
9. Analyze by immunoblotting and streptavidin blotting (*see* **Subheading 3.5.**).

3.4.2. Labeling in Living Cells

3.4.2.1. CELL CULTURE AND LABELING

1. Culture cells as above (*see* **Subheading 3.4.1.1.**).
2. Treat cells for the desired time with the desired apoptotic stimulus to induce apoptosis.
3. Add 1 μM O-methylated bEVD-aomk (*see* **Note 10**) to the culture media; replace into cell culture incubator for 1 h.
4. Collect cells by centrifugation at $500 \times g$ for 10 min at 4°C; discard supernatant.

5. Wash cells with 10 ml of cold PBS; harvest cells as above.
6. Transfer cells to a microfuge tube using 1 ml of cold PBS.
7. Drain cell pellet with a 27-gauge needle attached to a vacuum line.

3.4.2.2. Preparation of Extract

1. Resuspend cells into 100 μl of mRIPA buffer containing protease inhibitors (*see* **Note 9**) and caspase inhibitors (100 μM AcDEVD-CHO and 100 μM zVAD-fmk).
2. Incubate on ice for 20 min.
3. Clear lysate by centrifugation at 18,000 × *g* for 30 min at 4°C.
4. Transfer supernatant to a fresh tube.
5. Reserve 30 μl of sample for input control immunoblots.
6. Capture biotinylated proteins as described above (*see* **Subheading 3.4.1.3.**).

3.4.3. Labeling of Hypotonic Extract

ABPs are well suited to study caspase activation in hypotonic extracts. **Figure 7** presents a time-course of caspases activation in an extract programmed with cytochrome c. The bEVD-aomk is added after a lag period and allowed to react for 30 min. It strongly labeled large subunits of executioner caspases and mildly labeled proteins above 30 kDa that could correspond to initiator caspases.

1. Prepare and activate hypotonic extract as describe above (*see* **Subheading 3.2.1.**); omit AcDEVD-pNA substrate.
2. At the desired time point after addition of cytochrome c and dATP, add 5 μl of 10 μM bEVD-aomk probe to each reaction.
3. Further incubate reaction for 30 min at 37°C.
4. Add 54 μl of caspase dilution buffer to each reaction; thoroughly mix.
5. Reserve 30 μl of sample for input control immunoblots.
7. Perform pull-down as described above (*see* **Subheading 3.4.1.3.**).

3.4.4. Labeling of Recombinant Proteins

The labeling of recombinant proteins is done in a similar fashion as done for an enzymatic assay (*see* **Subheadings 3.1.1.** and **3.1.2.**) with the exception that the substrate is replaced with a molar excess of ABP. **Figure 8** was generated by incubating various active and inactive recombinant caspases (*see* figure) with bVAD-fmk ABP for 1 h in assay buffer as described in **Subheading 3.1**. Proteins were analyzed as in **Subheading 3.5** by SDS–PAGE and streptavidin blot.

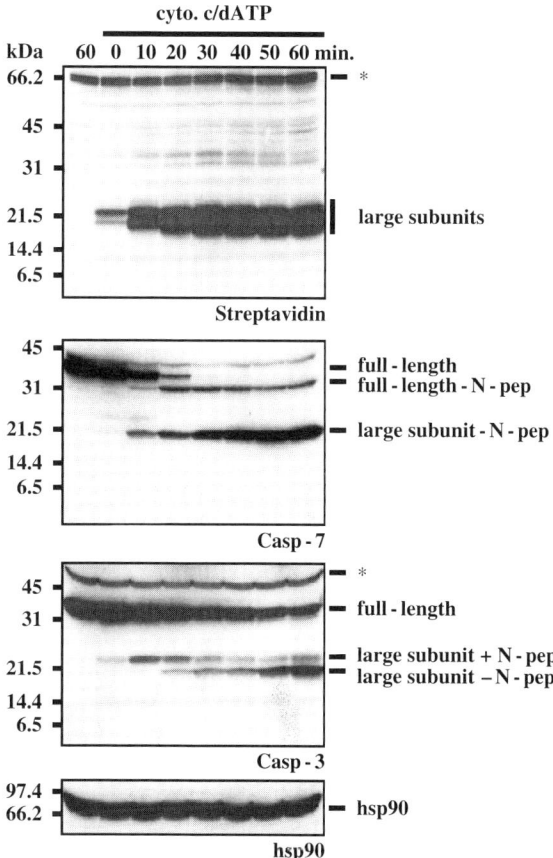

Fig. 7. A series of cytochrome c/dATP activation assays was set and 10 μM bEVD-acyloxymethyl ketone (bEVD-aomk) was added at the indicated time point. Assays were further incubated for 30 min. In the streptavidin blot, the asterisk indicates a endogenously biotinylated protein (carboxylase) and it indicates a non-specific band in the caspase-3 blot. Ensuing samples were analyzed using streptavidin–HRP and caspase-7 or caspase-3 antibodies (*see* **Subheading 3.5.**).

3.5. Analysis by Immunoblot and Streptavidin Blot

3.5.1. By Immunoblot

The gel apparatus system used is the one from CBS Scientific, Del Mar, CA, USA and gels are cast 10 at a time in a multigel-casting chamber using 0.75-mm spacers and combs. The volume of gel mix hence needs to be modified

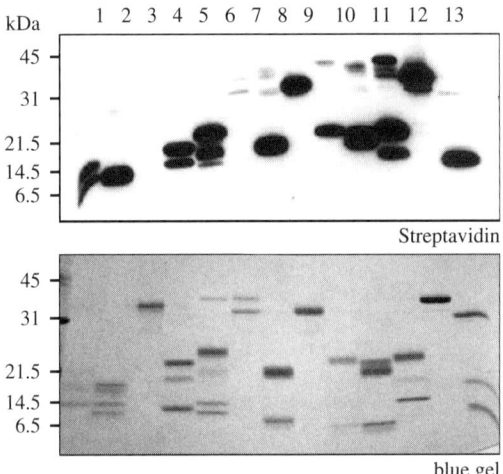

Fig. 8. Recombinant caspases (1 μM) were incubated for 1 h in their corresponding buffer (caspase buffer, caspase-3/6/7/14; high salt caspase buffer, caspase-2/8/9/10) in the presence of 10 μM b-Val-Ala-Asp-fluoromethyl ketone (bVAD-fmk) (20 μM for caspase-2). An aliquot of the protein was analyzed by SDS–PAGE or streptavidin–HRP blotting. Lanes: 1, ΔCARD-Casp-2; 2, Casp-3; 3, Casp-3 C285A; 4, Casp-6; 5, ΔN-Casp-7; 6, pro-Casp-7; 7, ΔDEDs-Casp-8; 8, uncleavable ΔDEDs-Casp-8; 9, ΔCARD-Casp-9 preparation 1; 10, ΔCARD-Casp-9 preparation 2; 11, ΔDEDs-Casp-10; 12, uncleavable ΔDEDs-Casp-10; 13, Casp-14.

according to the gel system used. We routinely use 8–18% acrylamide gradient gels in a 2-amino-2-methyl-1,3-propandiol (ammediol) buffer system. This buffer system provides good resolution over a great range of molecular weight and is particularly good for visualizing small proteins such as caspase subunits. But in principle, any SDS–PAGE system will work. The transfer is performed to polyvinylidene fluoride (PVDF) membrane using a Hoefer chamber for mini-gels using a CAPS buffer *(49)*.

3.5.1.1. Gel Preparation

1. Assemble the gel-casting chamber, connect to a gradient maker with a peristaltic pump in between, and set the pump to deliver 3.0 ml/min.
2. Prepare low and high acrylamide mixes without APS or TEMED (*see* **Subheading 2.2.** and **Note 11**).
3. Filter to degas using a 50-ml 0.22-μm membrane filter unit.
4. Add 20 ml of ethanol–LGB solution into the gel-casting chamber.
5. Add TEMED and 10% APS solution to each acrylamide mixes; mix gently.

6. Pour both solutions into the gradient maker so the pump draws the liquid from the low acrylamide solution first.
7. Start pump immediately and stop when the un-polymerized gel level is about 1 cm below where the bottom of a well would be.
8. Stop pump, close gel-casting chamber valve, and rinse tubing immediately with ddH$_2$O.
9. Let gel set for at least 4 h.
10. Disassemble the gel-casting unit, remove ethanol solution, and rinse gels with ddH$_2$O.
11. Store gels at 4°C in a tight container with a wet paper towel to keep moisture.
12. One hour before running the gel, pour stacking gel (*see* **Subheading 2.2.**) and set comb in place.
13. Rinse wells with UGB prior to loading samples.

3.5.1.2. SAMPLE PREPARATION

1. Add 60 mM DTT to 3× gel-loading buffer just before use (1 ml).
2. Add 0.5 volume of 3× gel-loading buffer to the sample to be analyzed (*see* **Note 12**).
3. Prepare 2 μg of molecular weight markers in 1× gel-loading buffer to the same final volume as that of the sample to be analyzed.
4. Boil samples for 3 min and chill on ice before loading.

3.5.1.3. RUNNING GELS

1. Fill the upper and lower chambers with 1× URB and 1× LRB, respectively.
2. Load 10–30 μl of samples per well; a similar volume of diluted molecular weight marker is loaded on each extremity of the gel to prevent too much band expansion.
3. Connect the gel chamber and apply 30 mA per gel until the bromophenol blue dye runs out of the gel.
4. Disassemble the gel and carefully remove the stacking gel.

3.5.1.4. TRANSFER

1. Cut PVDF membrane to 7 × 10 cm and soak in 50 ml of methanol for 5 s.
2. Soak membrane in 50 ml of transfer buffer for 10 min.
3. Soak gel in 100 ml of transfer buffer for 10 min.
4. Assemble transfer pads, filter paper, gel, and membrane into transfer cassette.
5. Transfer for 45 min at constant current of 0.4 A with cooling.
6. Disassemble cassette and transfer membrane into ddH$_2$O.
7. Stain membrane for 2 min in 10 ml Ponceau S solution.
8. Rinse with plenty of ddH$_2$O.
9. Mark molecular weight markers with a pencil.
10. Keep in ddH$_2$O.

3.5.1.5. IMMUNOBLOTTING

1. Prepare 50 ml of PBS-T-milk.
2. Incubate membrane in 35 ml of PBS-T-milk for 1 h at room temperature.
3. Store the remaining 15 ml of PBS-T-milk at 4°C for the secondary antibody.
4. Prepare 10 ml of PBS–BSA with primary antibody (*see* **Note 13**); keep on ice.
5. Rinse membrane three times with 20 ml PBS-T for 5 s.
6. Incubate membrane in antibody solution overnight at 4°C (*see* **Note 13**).
7. Rinse membrane three times with PBS-T for 5 s and three times for 15 min.
8. Incubate membrane in 15 ml of PBS-T-milk containing the secondary antibody (1:5000) for 2 h at room temperature.
9. Rinse membrane as above.

3.5.1.6. DEVELOPING

1. Rinse membrane in PBS; drain liquid.
2. Overlay the membrane with WestPico SuperSignal detection reagent; make sure the membrane is fully covered.
3. Incubate for 5 min.
4. Drain excess liquid.
5. Carefully wrap membrane in transparent sheets, and remove any air bubble by rolling a Pasteur pipette over the covered membrane.
6. Expose to film for 2 min and develop.
7. Readjust exposure time if necessary.

3.5.2. By Streptavidin Blot

The entire procedure to perform a streptavidin blot is identical to the one described above (*see* **Subheading 3.5.**) with the exception of the immunoblotting procedure (*see* below). Furthermore, because of the high sensitivity of biotin detection, less volume of sample is required; routinely one-fourth of the amount used for immunoblotting is sufficient.

1. Prepare 50 ml of PBS-T-iBlock.
2. Incubate the PVDF membrane in 35 ml of PBS-T-iBlock for 1 h at room temperature.
3. Prepare 15 ml of PBS-T-iBlock with 0.2 µg/ml of streptavidin–HRP; keep on ice.
4. Incubate membrane in diluted streptavidin–HRP overnight at 4°C (*see* **Note 13**).
5. Rinse membrane three times with PBS-T for 5 s and three times for 10 min.

Membrane is ready for development (*see* **Subheading 3.5.1.6.**).

4. Notes

1. P1 refers to the residue positioned N-terminus to the scissile bond cleaved by the protease and extends toward the N-terminus (position P2, P3, …) of the polypeptide, whereas prime positions (P1´, P2´, …) are C-terminus to the cleaved

peptide bond *(50)*. Those positions are matched by substrate-binding pockets receiving substrate residues (S1 or S1´ and so on).

2. With the preferred Afc-coupled substrates, the following concentration of caspases can be used but should be optimized for the fluorometer used: caspase-2 (Val-Asp-Val-Ala-Asp (VDVAD)), 20 nM; caspase-3 (DEVD), 0.1–1 nM; caspase-6 (Val-Glu-Ile-Asp (VEID) or DEVD), 10 nM; caspase-7 (DEVD), 5–10 nM; caspase-8 (Ile-Glu-Thr-Asp (IETD) or DEVD), 20–50 nM; caspase-9 (Leu-Glu-His-Asp (LEHD), 10–20 nM; and caspase-10 (IETD or DEVD), 40–50 nM.

3. Recombinant caspases can be obtained from several commercial sources or can be made in-house rather easily *(44,51)*. Because of the variability in purity from different suppliers, a critical aspect to consider is the usage of an active-site titrated caspase preparation instead of an enzyme concentration based solely on protein amount.

4. Initiator caspases are fully active only when dimerized. Because dimerization is a bimolecular reaction, the higher the concentration, the faster the reaction. Initiator caspases 8, 9, and 10 will fully dimerize at 100 nM in 1 M sodium citrate buffer over a 1-h period.

5. Continuous read kinetics has several advantages over end point kinetics. It allows the detection of substrate depletion that would otherwise lead to serious underestimation of the activity of fast reactions. It also permits the detection of enzyme instability displayed by a decreasing slope of product generation over time.

6. Because the enzyme concentration will be high in the uninhibited sample, substrate depletion may occur within minutes especially with caspase-3. It is thus critical to read samples as fast and as soon as possible. One possibility is to add substrate starting with wells containing the highest zVAD-fmk concentration and do one titration at a time. Also, the enzyme inhibitor mixtures can be diluted 10–50 times before addition of the reporter substrate.

7. Samples prepared in detergent extracts should not be frozen prior to performing the enzymatic assays because of further caspase activation and increased background; they should be analyzed as soon as possible.

8. The ABP bEVD-aomk probe is not commercially available. In most instances, it can be substituted by bVAD-fmk. The advantage of bEVD-aomk over bVAD-fmk is not only that aomk warhead is more specific than the fmk one and that the EVD sequence is better recognized by caspases but that the biotin affinity tag is spaced from the core of the ABP by a six-carbon spacer. This allows better capture of the labeled proteins.

9. If post-lysis proteolytic activity is a problem, the following protease inhibitors could be added to the lysis buffer without interfering with the assay: 5 mM EDTA, 1 mM 1,10-orthophenanthroline, 10 μM E-64, 10 μM leupeptin, 10 μM 3,4-dichloroisocoumarin, and 1 μM MG-132. However, iodoacetamide, *N*-methyl maleimide, or other cysteine-modifying agents must be avoided because caspases utilize a cysteine residue in catalysis.

10. Some non-O-methylated ABPs such as bEVD-aomk are cell permeable and can be used instead of O-methylated ABPs. The cell-penetrating activities of caspase ABPs have never been reported, and O-methylated Asp derivatives are generally used in the hope that they will penetrate cells better. However, in our hands, non-esterified probes do enter cells well enough for most applications.

11. The volume of gel mix to be used differs from gel system to another so it should be adjusted. For mini-gels, the resolving gel level should be 1 cm below the well.

12. If samples contain high salt, they should either be diluted at least 10-fold before preparation so the sodium citrate concentration is 100 mM or lower. Alternatively, proteins can be precipitated and recovered using acetone or trichloroacetic acid (TCA).

13. The amount of antibody to use and time of blotting vary greatly between antibodies and should be optimized for each analysis. As a starting point, using half the supplier's recommended concentration and overnight blotting should give reasonable results. Blotting overnight allows the use of the smallest amount of antibody possible without loss of signal or sensitivity.

Acknowledgments

We are grateful to the members of our laboratory who provided some of the results presented in this chapter and for carefully reviewing protocols.

References

1. Denault, J. B., and Salvesen, G. S. (2002) Caspases: keys in the ignition of cell death. *Chem Rev* **102,** 4489–500.

2. Fischer, U., Janicke, R. U., and Schulze-Osthoff, K. (2003) Many cuts to ruin: a comprehensive update of caspase substrates. *Cell Death Differ* **10,** 76–100.

3. Woo, E. J., Kim, Y. G., Kim, M. S., Han, W. D., Shin, S., Robinson, H., Park, S. Y., and Oh, B. H. (2004) Structural mechanism for inactivation and activation of CAD/DFF40 in the apoptotic pathway. *Mol Cell* **14,** 531–9.

4. Sakahira, H., Enari, M., and Nagata, S. (1998) Cleavage of CAD inhibitor in CAD activation and DNA degradation during apoptosis. *Nature* **391,** 96–9.

5. Liu, X., Zou, H., Widlak, P., Garrard, W., and Wang, X. (1999) Activation of the apoptotic endonuclease DFF40 (caspase-activated DNase or nuclease). Oligomerization and direct interaction with histone H1. *J Biol Chem* **274,** 13836–40.

6. Liu, X., Zou, H., Slaughter, C., and Wang, X. (1997) DFF, a heterodimeric protein that functions downstream of caspase-3 to trigger DNA fragmentation during apoptosis. *Cell* **89,** 175–84.

7. Enari, M., Sakahira, H., Yokoyama, H., Okawa, K., Iwamatsu, A., and Nagata, S. (1998) A caspase-activated DNase that degrades DNA during apoptosis, and its inhibitor ICAD. *Nature* **391,** 43–50.

8. Lazebnik, Y. A., Kaufmann, S. H., Desnoyers, S., Poirier, G. G., and Earnshaw, W. C. (1994) Cleavage of poly(ADP-ribose) polymerase by a proteinase with properties like ICE. *Nature* **371,** 346–7.

9. Luo, X., Budihardjo, I., Zou, H., Slaughter, C., and Wang, X. (1998) Bid, a Bcl2 interacting protein, mediates cytochrome c release from mitochondria in response to activation of cell surface death receptors. *Cell* **94,** 481–90.

10. Li, H., Zhu, H., Xu, C. J., and Yuan, J. (1998) Cleavage of BID by caspase 8 mediates the mitochondrial damage in the Fas pathway of apoptosis. *Cell* **94,** 491–501.

11. Cerretti, D. P., Kozlosky, C. J., Mosley, B., Nelson, N., Van Ness, K., Greenstreet, T. A., March, C. J., Kronheim, S. R., Druck, T., Cannizzaro, L. A., Huebner, K., and Black, R. A. (1992) Molecular cloning of the interleukin-1b converting enzyme. *Science* **256,** 97–100.

12. Thornberry, N. A., Bull, H. G., Calaycay, J. R., Chapman, K. T., Howard, A. D., Kostura, M. J., Miller, D. K., Molineaux, S. M., Weidner, J. R., Aunins, J., Elliston, K. O., Ayala, J. M., Casano, F. J., Chin, J., Ding, G. J. F., Egger, L. A., Gaffney, E. P., Limjuco, G., Palyha, O. C., Raju, S. M., Rolando, A. M., Salley, J. P., Yamin, T. T., and Tocci, M. J. (1992) A novel heterodimeric cysteine protease is required for interleukin-1beta processing in monocytes. *Nature* **356,** 768–74.

13. Alam, A., Cohen, L. Y., Aouad, S., and Sekaly, R. P. (1999) Early activation of caspases during T lymphocyte stimulation results in selective substrate cleavage in nonapoptotic cells. *J Exp Med* **190,** 1879–90.

14. Kennedy, N. J., Kataoka, T., Tschopp, J., and Budd, R. C. (1999) Caspase activation is required for T cell proliferation. *J Exp Med* **190,** 1891–6.

15. Algeciras-Schimnich, A., Barnhart, B. C., and Peter, M. E. (2002) Apoptosis-independent functions of killer caspases. *Curr Opin Cell Biol* **14,** 721–6.

16. Fuentes-Prior, P., and Salvesen, G. S. (2004) The protein structures that shape caspase activity, specificity, activation and inhibition. *Biochem J* **384,** 201–32.

17. Zhou, Q., and Salvesen, G. S. (1997) Activation of pro-caspase-7 by serine proteases includes a non-canonical specificity. *Biochem J* **324,** 361–64.

18. Boatright, K. M., Renatus, M., Scott, F. L., Sperandio, S., Shin, H., Pedersen, I., Ricci, J.-E., Edris, W. A., Sutherlin, D. P., Green, D. R., and Salvesen, G. S. (2003) A unified model for apical caspase activation. *Mol Cell* **11,** 529–41.

19. Talanian, R. V., Dang, L. C., Ferenz, C. R., Hackett, M. C., Mankovich, J. A., Welch, J. P., Wong, W. W., and Brady, K. D. (1996) Stability and oligomeric equilibria of refolded interleukin-1beta converting enzyme. *J Biol Chem* **271,** 21853–8.

20. Stennicke, H. R., Deveraux, Q. L., Humke, E. W., Reed, J. C., Dixit, V. M., and Salvesen, G. S. (1999) Caspase-9 can be activated without proteolytic processing. *J Biol Chem* **274,** 8359–62.

21. Cowling, V., and Downward, J. (2002) Caspase-6 is the direct activator of caspase-8 in the cytochrome c-induced apoptosis pathway: absolute requirement for removal of caspase-6 prodomain. *Cell Death Differ* **9**, 1046–56.

22. Pop, C., Timmer, J., Sperandio, S., and Salvesen, G. S. (2006) The apoptosome activates caspase-9 by dimerization. *Mol Cell* **22**, 269–75.

23. Cain, K., Bratton, S. B., Langlais, C., Walker, G., Brown, D. G., Sun, X. M., and Cohen, G. M. (2000) Apaf-1 oligomerizes into biologically active approximately 700-kDa and inactive approximately 1.4-MDa apoptosome complexes. *J Biol Chem* **275**, 6067–70.

24. Zou, H., Li, Y., Liu, X., and Wang, X. (1999) An APAF-1.cytochrome c multimeric complex is a functional apoptosome that activates procaspase-9. *J Biol Chem* **274**, 11549–56.

25. Yu, X., Acehan, D., Menetret, J. F., Booth, C. R., Ludtke, S. J., Riedl, S. J., Shi, Y., Wang, X., and Akey, C. W. (2005) A structure of the human apoptosome at 12.8 A resolution provides insights into this cell death platform. *Structure (Camb)* **13**, 1725–35.

26. Riedl, S. J., Li, W., Chao, Y., Schwarzenbacher, R., and Shi, Y. (2005) Structure of the apoptotic protease-activating factor 1 bound to ADP. *Nature* **434**, 926–33.

27. Peter, M. E., and Krammer, P. H. (2003) The CD95(APO-1/Fas) DISC and beyond. *Cell Death Differ* **10**, 26–35.

28. Denault, J. B., and Salvesen, G. S. (2003) Human caspase-7 activity and regulation by its N-terminal Peptide. *J Biol Chem* **278**, 34042–50.

29. Tenev, T., Zachariou, A., Wilson, R., Ditzel, M., and Meier, P. (2005) IAPs are functionally non-equivalent and regulate effector caspases through distinct mechanisms. *Nat Cell Biol* **7**, 70–7.

30. Scott, F. L., Denault, J. B., Riedl, S. J., Shin, H., Renatus, M., and Salvesen, G. S. (2005) XIAP inhibits caspase-3 and -7 using two binding sites: evolutionarily conserved mechanism of IAPs. *EMBO J* **24**, 645–55.

31. Riedl, S. J., Renatus, M., Schwarzenbacher, R., Zhou, Q., Sun, S., Fesik, S. W., Liddington, R. C., and Salvesen, G. S. (2001) Structural basis for the inhibition of caspase-3 by XIAP. *Cell* **104**, 791–800.

32. Vaux, D. L. (1999) Caspases and apoptosis – biology and terminology. *Cell Death Differ* **6**, 493–4.

33. Thornberry, N. A., Rano, T. A., Peterson, E. P., Rasper, D. M., Timkey, T., Garcia-Calvo, M., Houtzager, V. M., Nordstrom, P. A., Roy, S., Vaillancourt, J. P., Chapman, K. T., and Nicholson, D. W. (1997) A combinatorial approach defines specificities of members of the caspase family and granzyme B. Functional relationships established for key mediators of apoptosis. *J Biol Chem* **272**, 17907–11.

34. Stennicke, H. R., Renatus, M., Meldal, M., and Salvesen, G. S. (2000) Internally quenched fluorescent peptide substrates disclose the subsite preferences of human caspases 1, 3, 6, 7 and 8. *Biochem J* **350**, 563–68.

35. Deveraux, Q., Takahashi, R., Salvesen, G. S., and Reed, J. C. (1997) X-linked IAP is a direct inhibitor of cell death proteases. *Nature* **388,** 300–04.
36. Takahashi, R., Deveraux, Q., Tamm, I., Welsh, K., Assa-Munt, N., Salvesen, G. S., and Reed, J. C. (1998) A single BIR domain of XIAP sufficient for inhibiting caspases. *J Biol Chem* **273,** 7787–90.
37. Schimmer, A. D., Welsh, K., Pinilla, C., Wang, Z., Krajewska, M., Bonneau, M. J., Pedersen, I. M., Kitada, S., Scott, F. L., Bailly-Maitre, B., Glinsky, G., Scudiero, D., Sausville, E., Salvesen, G., Nefzi, A., Ostresh, J. M., Houghten, R. A., and Reed, J. C. (2004) Small-molecule antagonists of apoptosis suppressor XIAP exhibit broad antitumor activity. *Cancer Cell* **5,** 25–35.
38. Schimmer, A. D., Dalili, S., Batey, R. A., and Riedl, S. J. (2006) Targeting XIAP for the treatment of malignancy. *Cell Death Differ* **13,** 179–88.
39. Kato, D., Boatright, K. M., Berger, A. B., Nazif, T., Blum, G., Ryan, C., Chehade, K. A. H., Salvesen, G. S., and Bogyo, M. (2005) Activity-based probes that target diverse cysteine protease families. *Nat Chem Biol* **1,** 33–38.
40. Nicholson, D. W., Ali, A., Thornberry, N. A., Vaillancourt, J. P., Ding, C. K., Gallant, M., Gareau, Y., Griffin, P. R., Labelle, M., Lazebnik, Y. A., et al. (1995) Identification and inhibition of the ICE/CED-3 protease necessary for mammalian apoptosis. *Nature* **376,** 37–43.
41. Berger, A. B., Witte, M., Denault, J. B., Sagadhiani, A. M., Sexton, K. M. B., Salvesen, G. S., and Bogyo, M. (2006) Identification of early intermediates of caspase activation during intrinsic apoptosis using selective inhibitors and activity based probes. *Mol Cell* **23,** 509–21.
42. Donepudi, M., Mac Sweeney, A., Briand, C., and Gruetter, M. G. (2003) Insights into the regulatory mechanism for caspase-8 activation. *Mol Cell* **11,** 543–49.
43. Edelhoch, H. (1967) Spectroscopic determination of tryptophan and tyrosine in proteins. *Biochemistry* **6,** 1948–54.
44. Stennicke, H. R., and Salvesen, G. S. (1999) Caspases: preparation and characterization. *Methods* **17,** 313–9.
45. Li, P., Nijhawan, D., Budihardjo, I., Srinivasula, S. M., Ahmad, M., Alnemri, E. S., and Wang, X. (1997) Cytochrome c and dATP-dependent formation of Apaf-1/caspase-9 complex initiates an apoptotic protease cascade. *Cell* **91,** 479–89.
46. Ellerby, H. M., Martin, S. J., Ellerby, L. M., Naiem, S. S., Rabizadeh, S., Salvesen, G. S., Casiano, C. A., Cashman, N. R., Green, D. R., and Bredesen, D. E. (1997) Establishment of a cell-free system of neuronal apoptosis: comparison of premitochondrial, mitochondrial, and postmitochondrial phases. *J Neurosci* **17,** 6165–78.
47. Denault, J. B., Békés, M., Scott, F. L., Sexton, K. M. B., Bogyo, M., and Salvesen, G. S. (2006) Engineered hybrid dimers: tracking the activation pathway of caspase-7. *Mol Cell* **23,** 523–33.
48. Stennicke, H. R., Jurgensmeier, J. M., Shin, H., Deveraux, Q., Wolf, B. B., Yang, X., Zhou, Q., Ellerby, H. M., Ellerby, L. M., Bredesen, D., Green, D. R.,

Reed, J. C., Froelich, C. J., and Salvesen, G. S. (1998) Pro-caspase-3 is a major physiologic target of caspase-8. *J Biol Chem* **273,** 27084–90.

49. Matsudaira, P. (1987) Sequence from picomole quantities of proteins electroblotted onto polyvinylidene difluoride membranes. *J Biol Chem* **262,** 10035–38.
50. Schecter, I., and Berger, M. (1967) On the size of the active site in proteases. *Biochem Biophys Res Commun* **27,** 157–62.
51. Denault, J. -B., and Salvesen, G. S. (2002) Unit 21.13: Expression, purification and characterization of caspases, Current Protocols in Protein Sciences, John Wiley & Sons, Academic Press.

16

Biochemical Analysis of the Native TRAIL Death-Inducing Signaling Complex

Henning Walczak and Tobias L. Haas

Summary

The extrinsic apoptosis pathway is activated when certain members of the tumor necrosis factor (TNF) receptor superfamily (TNFRSF) are oligomerized by their cognate ligands that are members of the TNF superfamily (TNFSF). The apoptosis-inducing capacity of a member of the TNFRSF relies on the presence of a death domain (DD) in the intracellular portion of the receptor protein. Such receptors are also referred to as death receptors. Binding of a TNFSF ligand to a TNFRSF receptor that is expressed on the surface of a cell results in the formation of a receptor proximal protein complex. This protein complex is the platform for further signaling events within the cell. In case of death receptors like TNF-related apoptosis-inducing ligand receptor 1 (TRAIL-R1/DR4), TRAIL-R2 (KILLER/APO-2/DR5/TRICK), CD95 (Fas, APO-1), or TNF receptor 1 (TNF-R1), this complex is termed death-inducing signaling complex (DISC). The compositions of the various DISCs have been intensively studied in the last 12 years. For the CD95 and the TRAIL-R1/R2 DISCs, it is now clear that the adaptor protein Fas-associated DD protein (FADD) forms part of these complexes and is necessary for recruitment of the pro-apoptotic signaling molecules caspase-8 and caspase-10. Recruitment of these proteases allows for their activation at the DISC and subsequent induction of apoptosis. The caspase-8 homologous cellular FLICE-like inhibitory protein (cFLIP) can also be recruited to the DISC. cFLIP acts as an anti-apoptotic regulator by interfering with activation of caspases 8 and 10 at the DISC. Interestingly, treatment of TRAIL-resistant tumor cells with conventional chemotherapeutic drugs or with proteasome inhibitors renders these cells sensitive for TRAIL-induced apoptosis. By applying the methodology of the biochemical analysis of the TRAIL DISC described here, we were able to show that this sensitization is mainly due to changes in the biochemical composition of the DISC as the apoptosis-initiating protein complex of the extrinsic pathway.

Key Words: TRAIL; apoptosis; DISC; immunoprecipitation; caspase.

From: *Methods in Molecular Biology, vol. 414: Apoptosis and Cancer*
Edited by: G. Mor and A. B. Alvero © Humana Press Inc., Totowa, NJ

1. Introduction

Programmed cell death by apoptosis is a morphologically and biochemically distinct event allowing for tissue homeostasis and removal of old, superfluous, and potentially dangerous cells. Apoptosis plays an important role during embryonic development in the immune system and in tissue homeostasis. Two main apoptosis signaling pathways are known. Various forms of cellular stress result in activation of the "intrinsic" or mitochondrial apoptosis pathway. The "extrinsic" pathway is activated by death ligand-induced oligomerization of death receptors. In both cases, "initiator" caspases are activated. This activation occurs at high molecular weight protein complexes. Depending on the stimulus, this complex is either the apoptosome for the intrinsic pathway or the death-inducing signaling complex (DISC) for the extrinsic pathway. The apoptosome serves as the activation platform for caspase-9 and the DISC as the activation platform for caspase-8 and caspase-10. Together with caspase-2, these caspases are the so-called "initiator" caspases. Initiator caspases are characterized by long prodomains. These domains are required for their recruitment to and activation at the above-mentioned protein complexes. The activation of initiator caspases is needed to directly or indirectly activate the "effector" caspases 3, 6, and 7. These caspases are then responsible for the demolition phase of apoptosis. The sequential and ordered activation of initiator and effector caspases allows for controlled cellular destruction. The proteolytic cascade triggered by initiator caspase activation results in the typical morphological hallmarks of apoptosis. These include nuclear fragmentation, chromatin condensation, DNA fragmentation, cell shrinkage, and exposure of phosphatidyl serine on the outer leaflet of the plasma membrane. The latter event allows for the recognition of a dying cell by phagocytes.

The extrinsic pathway is activated by death receptors belonging to the tumor necrosis factor (TNF) receptor superfamily (TNFRSF). The death receptor subfamily includes the two death receptors for the TNF-related apoptosis-inducing ligand (TRAIL), that is, TRAIL-R1 (DR4) and TRAIL-R2 (DR5/APO-2/Killer/TRICK2), CD95 (APO 1/Fas), the TNF receptor 1 (TNFR-1; also known as p55 TNFR), and TNF related apoptosis mediating protein (TRAMP/DR3). Upon crosslinking of death receptors by their corresponding ligands or agonistic antibodies, a caspase cascade that triggers the controlled death of the cell by apoptosis is activated. The TRAIL/TRAIL-R system is particularly attractive for cancer researchers because of its apparent specificity for killing of transformed cells while sparing normal cells (*1*). It will be the key to the exploitation of the full therapeutic potential of agonists of the two

TRAIL death receptors in the treatment of cancer to understand the biochemical basis of the tumor selectivity of TRAIL *(2)*.

TRAIL, also known as Apo2L, was first described in 1995 *(3)*. It was identified due to its homology to the ligand of CD95 (CD95L/FasL) (28% identity) and TNF-α (23% identity). TRAIL is expressed as a type II transmembrane protein consisting of 281 amino acids with a short cytoplasmic N-terminus and a long C-terminal extracellular receptor-binding domain. Like TNF-α and CD95L, the carboxy-terminal extracellular part of TRAIL can be cleaved off from the cell membrane by specific metalloproteases under certain conditions. In solution, soluble TRAIL forms a homotrimer, which is stabilized by a cysteine residue at position 230 that coordinates with a divalent zinc ion *(4)*. Soluble untagged recombinant trimerized TRAIL did not induce significant cell death at doses up to 100 μg/ml in human or cynomolgus monkey hepatocytes or in human keratinocytes, whereas certain aggregated or antibody-crosslinked forms of TRAIL may be toxic for these cells *(5)*. This difference might be due to differential stimulation of the TRAIL receptors by forming high- versus low-order oligomers of the ligand or simply by different stimulation capabilities for different cellular TRAIL receptors exerted by differently prepared ligands. At any rate, the fact that certain forms of TRAIL as well as certain TRAIL death receptor-specific agonistic antibodies exerted a cytotoxic effect on several tumor cell lines, whereas sparing normal cells raised the hope to use such TRAIL receptor agonists as novel tumor cell-selective anti-cancer drugs.

A total of five different cellular receptors bind to TRAIL. These are TRAIL-R1 (DR4), TRAIL-R2 (DR5/APO-2/Killer/TRICK2), TRAIL-R3 (DcR1/TRID/LIT), TRAIL-R4 (DcR2/TRUNDD), and osteoprotegerin (OPG).

TRAIL-R1 (DR4) was cloned by sequence comparisons between expressed sequence tags (ESTs) in public and private sequence libraries and known sequences of other death receptors *(6)*. Interestingly, one part of its cytoplasmic region showed significant homology to a domain found in CD95 and TNF-R1. This domain, the so-called death domain (DD), is critical for apoptosis induction and characteristic for all death receptors of the TNFRSF. TRAIL-R1 shares another characteristic feature with TNF-R1 and CD95, namely two complete cysteine-rich domains (CRDs) in the extracellular part of the protein. These are important for the binding of the respective death ligands.

Three different ways led to the identification of TRAIL-R2. We biochemically purified TRAIL-R2 and subsequently cloned the corresponding cDNA using TRAIL-based affinity chromatography followed by 2D gel electrophoresis, tandem mass spectrometric analysis of peptides derived from specific protein spots, and cloning of the TRAIL-R2 cDNA with peptide-derived oligonucleotides *(7)*. Coming from a totally different angle, Wu et al. *(8)* identified TRAIL-R2 as a p53-inducible gene. Lastly, TRAIL-R2 was identified by various groups in EST-based searches with the TRAIL-R1 sequence taking advantage of the 58% sequence homology that TRAIL-R2 shares with TRAIL-R1 *(9–13)*.

TRAIL-R1 and TRAIL-R2 are both capable of transmitting an apoptotic signal upon ligand binding. Interestingly, there are two alternative splice forms of TRAIL-R2, TRAIL-R2a (TRICK2A), and TRAIL-R2b (TRICK2B) but not of TRAIL-R1 *(11)*. The two forms of TRAIL-R2 differ in a 23 amino acid stretch between the transmembrane domain and the start of the first CRD, present in TRAIL-R2b and absent in TRAIL-R2a. Regarding functionality, so far no differences have been reported for the two isoforms of TRAIL-R2 *(14)*.

EST database searches and cDNA bank screens resulted in the identi-fication of two additional membrane-bound receptors for TRAIL, namely TRAIL-R3 (DcR1/TRID/LIT) and TRAIL-R4 (DcR2/TRUNDD). These two non-apoptosis-inducing TRAIL receptors, also referred to as "decoy receptors," are also capable of binding TRAIL, but unlike TRAIL-R1 and TRAIL-R2, they cannot transduce an apoptotic signal because they lack a functional DD.

TRAIL-R3 is a GPI-anchored molecule with high homology to the extracel-lular domain of TRAIL-R1 and TRAIL-R2. The transcripts of TRAIL-R3 are predominantly found on peripheral blood lymphocytes (PBLs). The signaling capacity of this receptor seems to be minimal because of the complete lack of an intracellular domain. A potential decoy function on TRAIL has been suggested but has only been shown in overexpression studies so far. High-ectopic expression of this receptor protected TRAIL-sensitive cells from under-going apoptosis upon TRAIL treatment, whereas artificial removal of the GPI anchor by phospholipases resensitized these cells for TRAIL-induced apoptosis. Specific upregulation of TRAIL-R3 was shown in tumors of the gastrointestinal tract. However, a TRAIL decoy function of TRAIL-R3 still awaits to be proven in a physiological setting *(14)*.

TRAIL-R4 was also found by homology search, comparing the sequence of TRAIL-R1 and TRAIL-R2 with EST databases (58–70% homology), and by performing a cDNA screen with the TRAIL-R3 sequence. TRAIL-R4 is widely expressed and not able to signal apoptosis due to the fact that it contains only an incomplete DD. Like TRAIL-R3, it can act as a "decoy receptor" upon overexpression, but again, this function has not yet been confirmed in a physiological setting. Overexpression of TRAIL-R4 results in the activation of nuclear factor κB (NF-κB), which upregulates several anti-apoptotic genes, resulting in protection from TRAIL-induced cell death *(15)*.

The last human TRAIL receptor identified is OPG. OPG is a soluble extracellular protein and best known for its activity as a regulator of the development and activation of osteoclasts in bone remodeling. OPG ligand (also called TRANCE or RANKL) binds to receptor activator of NF-κB (RANK) and stimulates osteoclastogenesis. In vitro, the affinity of OPG for TRAIL was shown to be about 10-fold lower than that of TRAIL-R1, TRAIL-R2, and TRAIL-R3 at physiological temperatures (37°C). Taken together, it is still unclear whether the TRAIL–OPG interaction is physiologically relevant or not *(14)*.

Signal transduction through most TNFRSF members is usually initiated by ligand-induced recruitment of cytosolic signaling molecules to the intracellular domain of the receptor. Protein complexes formed in this way contain adapter proteins that recruit effector proteins that are activated following recruitment. Upon stimulation of cells with TRAIL, the apoptosis-inducing signal is transmitted into the cell through the oligomerization of the TRAIL receptors TRAIL-R1 and/or TRAIL-R2.

Immunoprecipitation of the native TRAIL DISC, described in detail in this review, led to the identification of Fas-associated DD protein (FADD, also known as MORT1) as the adapter protein that was necessary for recruitment of procaspase-8 to the TRAIL-R1 and the TRAIL-R2 signaling complexes *(16–18)*. FADD also turned out to be the adapter protein necessary for the recruitment of procaspase-10 to ligand-crosslinked TRAIL-R1, TRAIL-R2, and CD95 *(19,20)*.

FADD is a small adapter protein consisting of 208 amino acid residues. It comprises two functionally and structurally distinct but related protein–protein interaction domains. According to the current model, the DD of FADD directly interacts with the DD of CD95, TRAIL-R1, or TRAIL-R2. Through its second functional domain, the death effector domain (DED), FADD recruits procaspase-8 and procaspase-10 to the protein complex. This model of direct interaction between the DDs of receptor and FADD was strengthened by the solution structures of isolated FADD and CD95 DDs. It suggests that the interaction takes place on a charged surface patch. However, recent reverse two-hybrid screens identified a series of point mutations in the DED of FADD abolishing the interaction with the CD95 DD indicating a potential role of the FADD DED in receptor interaction *(21)*.

Once recruited to the DISC, procaspase-8 is auto-catalytically activated. The active enzyme initiates apoptosis in TRAIL-sensitive cells by cleaving specific substrates like the pro-apoptotic of the Bcl-2 family member BID and the effector caspase-3. Caspase-3 is thereby activated and in turn cleaves a variety of cellular proteins, among them PARP, lamins, and cytokeratins. The massive proteolytic activity exerted by caspases leads to the characteristic apoptotic phenotype of the dying cell.

Native TRAIL and CD95 DISC analyses showed that besides procaspase-8, procaspase-10 also was recruited to and cleaved at these complexes *(19–22)*. However, the importance of caspase-10 as an initiator caspase in TRAIL-induced apoptosis is controversial. In two earlier studies, transient overexpression of caspase-10 could substitute for caspase-8 with respect to apoptosis induction. Yet recently, it was shown that this is not the case under more moderate, stable overexpression conditions. On the contrary, lymphocytes obtained from autoimmune lymphoproliferative syndrome (ALPS) type II patients that lack functional caspase-10 are less sensitive to TRAIL-induced apoptosis *(23)*. Also, in almost 50% of cancer cell lines investigated (mainly lung and breast cancer lines as well as some lymphoid cell lines), the protein level of procaspase-10 is reduced. This appears to be more common than the loss of caspase-8 *(5)*. Thus, loss of functional caspase-10 may be of physiological and pathological importance.

Both, procaspase-8 and procaspase-10 are expressed in different splice variants as single-chain inactive zymogens. The two major isoforms of procaspase-8, 8a and 8b, only differ in the linker region. Procaspase-10 is expressed on the protein level in at least three isoforms, 10a, 10c, and 10d. The 10c isoform consists only of the tandem DED and is catalytically inactive. Procaspases 10a and 10d, just like procaspases 8a and 8b, are recruited to the receptor complex, processed, and released as active enzymes *(19,20)*.

The cellular FLICE-like inhibitory protein (cFLIP) (FLICE is the former name for caspase-8) is a regulatory component of both the CD95 and the TRAIL DISC. By now, there are three different splice variants of cFLIP known to be expressed as polypeptides. $cFLIP_s$ (short form) and the recently characterized $cFLIP_R$ consist only of the linker and the tandem DEDs, just like procaspase-10c *(24)*. The long form of cFLIP, $cFLIP_L$, has an additional C-terminal caspase-like domain but lacks protease activity because the catalytically active cysteine is replaced by a tyrosine.

The cFLIP proteins are efficiently recruited to the TRAIL and CD95 DISC where $cFLIP_L$ is rapidly processed like the procaspases 8 and 10. The exact function of cFLIP in the complex is not unraveled to date. There are strong indications that $cFLIP_s$ has anti-apoptotic function. Recently, the property of $cFLIP_L$ as a solely anti-apoptotic regulator at the DISC has been challenged. The controversial outcomes of the experiments are highly depending on the experimental system used: studies in which high-stable overexpression was achieved uniformly showed that all three forms of cFLIP are potent inhibitors of TRAIL- and CD95L-induced apoptosis. Ectopic expression of $cFLIP_s$ led to markedly decreased recruitment of procaspase-8 to and activation at the CD95 DISC. Overexpression of $cFLIP_L$, on the contrary, resulted in the recruitment of both procaspase-8 and $cFLIP_L$ to the CD95 receptor complex. However, in this case, caspase-8 was only partially processed, and the cleaved intermediates remained bound to the DISC *(25)*. These results are in line with the findings that siRNA-mediated downregulation of cFLIP sensitized TRAIL-resistant cell lines for TRAIL-induced apoptosis *(26)*. Using targeted downregulation of either $cFLIP_L$ or $cFLIP_s$, it was recently shown that the anti-apoptotic role of $cFLIP_L$ is even more prominent than that of $cFLIP_S$ *(27)*. Data obtained on mouse embryonic fibroblasts deficient for cFLIP support the function of cFLIP as an inhibitor of caspase-8 activation, as these cells are more sensitive to death receptor-induced apoptosis. Also noteworthy is the finding that the expression of cFLIP correlates strongly with malignant potential of several cancers like hepatocellular carcinoma, colon adenocarcinoma, and melanoma *(28)*. On the contrary, it was observed that moderate ectopic $cFLIP_L$ expression could promote initial caspase-8 processing and enhance CD95-induced apoptosis. A possible explanation is the activation of caspase-8 by the caspase-like domain of cFLIPL through heterodimerization. This event may trap the activated but incompletely processed caspase-8 in the DISC, allowing for ongoing cleavage of DISC-proximal substrates *(29)*.

Phosphoprotein enriched in diabetes/phosphoprotein enriched in astrocytes-15 (PED/PEA 15) is another DED-containing protein that is recruited to the TRAIL DISC. Comparing the receptor complex of TRAIL-sensitive with TRAIL-resistant glioma cell lines, Xiao et al. found PED/PEA-15 only in the DISC of TRAIL resistant cells *(30)*. The differential recruitment among the various cell lines seemed to depend on the phosphorylation status of PEA-15 as recruitment of the double-phosphorylated form was only detectable in the TRAIL-resistant cell lines *(30)*. In addition, high endogenous expression of PED/PEA-15 protected primitive neuronal cells from death receptor-induced apoptosis. Recently, Ricci-Vitiani et al. *(31)* could demonstrate how in this system that downregulation of PEA-15 by siRNA sensitized these cells to apoptosis mediated by inflammatory cytokines and death receptors.

Analysis of the native TRAIL DISC was crucial for identification of factors that play an important role in TRAIL-induced apoptosis and for dissection of the underlying biochemical mechanisms. However, we are far from understanding all details of the initiation process of apoptosis induction by the DISC.

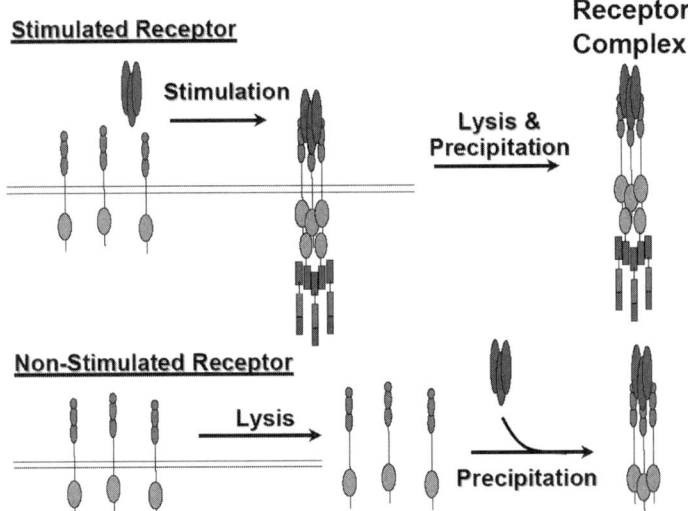

Fig. 1. Scheme of a death-inducing signaling complex (DISC) immunoprecipitation. For isolation of a stimulated receptor complex, the cells are incubated with a FLAG-tagged tumor necrosis factor (TNF) family member for 5–30 min. After washing away, unbound ligand cells are lysed in a physiological buffer by adding Triton X100. After lysis, the DISC is immunoprecipitated with antibodies against the N-terminal FLAG tag of the recombinant ligand. For the unstimulated control, cells are first lysed, and then the FLAG-tagged ligand is added, which binds to the receptor in the lysate. The receptor/ligand complexes are then precipitated by the addition of anti-FLAG antibody antibodies. These non-stimulated receptor–ligand complexes do not contain the stimulation-dependent DISC components.

Particularly, the biochemical analysis of regulatory DISC components will be an important focus in the future. Here, we provide a detailed description of the methodology we have successfully applied to identify and analyze core components of the native TRAIL DISC (*see* **Fig. 1**).

2. Materials

2.1. Equipment

1. 150 cm^2 tissue-culture flasks (Falcon, BD Biosciences, San Jose, CA, USA).
2. 75 cm^2 tissue-culture flasks (Falcon).
3. Cell scrapers (Falcon).
4. 50- and 15-ml conical tubes (BD Biosciences, San Jose, CA, USA).
5. 1.5-ml Eppendorf tubes.
6. Nitrocellulose membranes (Amersham Biosciences, Uppsala, Sweden).
7. Precast PAA Gel (NuPAGE 4–12%, Invitrogen, Carlsbad, CA, USA).

8. Mini-gel chamber + blotting unit (Invitrogen).
9. FACS Calibur (BD Biosciences).

2.2. Reagents

1. Phosphate-buffered saline (PBS).
2. Dulbecco's modified Eagle's medium (DMEM) + glutamax (Invitrogen).
3. 10× trypsin/EDTA (Invitrogen).
4. RPMI + glutamax (Invitrogen).
5. FBS (HyClone, Thermo Fisher, Waltham, MA, USA).
6. Fat-free powdered milk.
7. Tween 20 (GERBU, Gaiberg, Germany).
8. Propidium iodide (PI) (Sigma, Schnelldorf, Germany).
9. Bicinchoninic acid (BCA) assay (Pierce, Rockford, IL, USA).
10. Protease inhibitor cocktail (Roche Applied Science, Mannheim, Germany).
11. Streptavidin–phycoerythrin (SA-PE) or SA-FITC (BD Biosciences).
12. 0.2 g/l EDTA in ddH$_2$O (0.22-μm filtered).
13. FACS buffer: PBS + 5% FBS (prepare freshly before use).
14. Flow cytometry antibodies:

    a. HS101 (TRAIL-R1) (Axxora, San Diego, CA, USA).
    b. HS201 (TRAIL-R2) (Axxora).
    c. HS301 (TRAIL-R3) (Axxora).
    d. HS401 (TRAIL-R4) (Axxora).
    e. Anti-APO-1 (IgG1) (Axxora).
    f. Unspecific IgG1 isotype control (Southern Biotechnology, Birmingham, Alabama, USA).
    g. Biotinylated anti-mouse IgG1 (Southern Biotechnology).

15. Recombinant FLAG-TRAIL (Axxora).
16. Recombinant FLAG-CD95L (Axxora).
17. M2-antibody (anti-FLAG) (Sigma).
18. Protein-G sepharose (Amersham, Biosciences, Uppsala, Sweden).
19. Triton X100 (Sigma).
20. Supersignal West DURA ECL Solution (Pierce).
21. Lysis buffer: 30 mM Tris/HCl (pH 7.4) containing 150 mM NaCl, 5 mM KCl, 10% glycerol, 2 mM EDTA.
22. 4× LDS (lithium dodecyl sulfate) protein loading buffer (Invitrogen).
23. 500 mM TCEP (tris(2-carboxyethyl)phosphine) (Pierce, Rockford, IL, USA).
24. Antibodies for western blot:

    a. Anti-FADD (a66-2) (BD Biosciences, San Jose, CA, USA).
    b. Anti-FADD (clone 1) (BD Biosciences, San Jose, CA, USA).
    c. Anti-caspase-8 (C15) (Axxora).
    d. Anti-FLIP (NF6) (Axxora).
    e. Anti-TRAIL-R1 (CT) (Axxora).

    f. Anti-TRAIL-R2 (D3938) (Sigma).

    g. Anti-CD95 (C20) (Santa Cruz Biotechnology, Santa Cruz, CA, USA).

25. HRP-conjugated antibodies for western blot:

    a. Goat anti-mouse IgG1-HRP (Southern Biotechnology).

    b. Goat anti-mouse IgG2b-HRP (Southern Biotechnology).

    c. Goat anti-rabbit IgG (H+L)-HRP (Southern Biotechnology).

26. Stripping buffer for western blot: 50 mM glycine, pH 2.3.
27. Enhanced chemiluminescence (ECL) solution West DURA (Pierce).
28. 20x MOPS (3-(N-Morpholino)propanesulfonic acid) running buffer (Invitrogen, Carlsbad, CA, USA)

3. Methods

The methods are divided into (i) the cell culture of hepatocellular carcinoma (HCC) and leukemic cell lines (e.g., HepG2 and BJAB), (ii) the flow cytometric determination of the receptors, (iii) the DISC isolation by affinity purification, and (iv) the biochemical analysis of the complex by western blot.

3.1. Cell Culture

3.1.1. HCC Culture (e.g., HepG2 ATCC HB-8065)

Two days before DISC isolation, the HCC cells have to be seeded into two 150 cm^2 flasks with 6×10^6 cells per flask for DISC analysis and one 75 cm^2 flask with 3×10^6 cells for flow cytometric receptor analysis. All steps have to be carried out under sterile cell-culture conditions.

1. Wash the adherent cells with sterile $1\times$ PBS.
2. Trypsinize the cells with 5 ml prewarmed $1\times$ trypsin (stock diluted in $1\times$ PBS).
3. Incubate the cells with the trypsin/EDTA solution for 10 min at 37°C.
4. Stop the trypsination with 1 volume prewarmed DMEM + 10% FBS (the FBS will stop the trypsin).
5. Transfer the cell suspension into a 50-ml Falcon tube.
6. Count the cells.
7. Spin down and resuspend the cells in prewarmed DMEM + 10% FBS.
8. Seed 6×10^6 cells in each 150 cm^2 flask and 3×10^6 cells in the 75 cm^2 flask and fill to 25 ml (12.5 ml with prewarmed DMEM + 10% FBS, respectively).
9. Culture at 37°C and 10% CO2 for 48 h (24 h, respectively).

3.1.2. BJAB (Human Burkitt's Lymphoma Line) Culture

1. Count the suspension cells.
2. Spin down and resuspend the cells in prewarmed RPMI + 10% FBS.

3. Seed 2×10^7 cells in a 75 cm^2 flask for DISC analysis and 1×10^7 cells in a 25 cm^2 flask for FACS analysis and fill to 50 ml (12.5 ml with prewarmed RPMI + 10% FBS, respectively).
4. Culture at 37°C and 5% CO$_2$ for 48 h (24 h, respectively).

3.2. Flow Cytometric Determination of Receptor Expression

Expression of a given apoptosis-inducing receptor on the cell surface is a prerequisite for successful DISC isolation after incubation with its cognate ligand. In general, the expression level of the receptors correlates with the amount of isolated DISC. Therefore, one should make sure that the death receptors are expressed on the surface of the cells that are going to be used for DISC analysis. Receptor surface expression can most readily be determined by flow cytometry that is described in detail in this chapter.

3.2.1. Before Starting

1. Prepare FACS buffer (PBS + 5% FBS) and cool it on ice.
2. Warm the 0.2 g/l EDTA.
3. Prepare the antibody dilutions (each 10 μg/ml in ice-cold FACS buffer).
4. Prepare the SA-PE or SA-FITC dilutions (5 μg/ml in ice-cold FACS buffer, add 1 μg/ml PI and protect from light).

3.2.2. Preparation of the Cells for the Staining

1. Wash the adherent cells one time with 10 ml PBS per 75 cm^2 flask.
2. Incubate the cells with prewarmed 0.2 g/l EDTA for 5–10 min at 37°C.
3. Detach the cells by gently pipetting up and down.
4. Collect the cells in a 50-ml Falcon tube and count the cells.
5. Spin down at 1200–1500 g.
6. Resuspend the cells from 5×10^5 cells/ml to 1×10^6 cells/ml in ice-cold FACS buffer (same holds true for suspension cell lines).
7. Divide the cells into eight 1-ml Eppendorf tubes.

3.2.3. Staining Procedure (see **Table 1**)

1. Spin down the cells at 1200 g for 1 min in a precooled table-top centrifuge.
2. Remove the supernatant carefully.
3. Resuspend the cells in 50–100 μl primary antibody dilution (antibody diluted in ice-cold FACS buffer). For auto and PI-only control, add 50–100 μl FACS buffer.
4. Incubate for 20 min on ice.
5. Wash the cells:

   a. Add 1 ml ice-cold FACS buffer.
   b. Spin down for 1 min at 1200 g at 4°C.

Table 1
Recommended Staining Scheme

Sample	First antibody (10 μg/ml)	Second antibody (10 μg/ml)	SA-PE or SA-FITC	PI
Autofluorescence	–	–	–	–
PI	–	–	–	1 μg/ml
Isotype control	Mouse IgG1	Biotinylated goat anti-mouse IgG1	5 μg/ml	+
TRAIL-R1	HS101	Biotinylated goat anti-mouse IgG1	+	+
TRAIL-R2	HS201	Biotinylated goat anti-mouse IgG1	+	+
TRAIL-R3	HS301	Biotinylated goat anti-mouse IgG1	+	+
TRAIL-R4	HS401	+	+	+
Anti-APO-1	Anti-APO-1 (IgG1)	+	+	+

PE, phycoerythrin; PI, propidium iodide; SA, streptavidin; TRAIL-R1, tumor necrosis factor-related apoptosis-inducing ligand.

   c. Take off the supernatant.
   d. Resuspend again in 1 ml ice-cold FACS buffer (gently by pipetting).
   e. Spin down for 1 min at 1200 g at 4°C.
   f. Take off the supernatant carefully.

6. Add the 50–100 μl second antibody dilution (10 μg/ml diluted in ice-cold FACS buffer). For auto and PI-only control, add 50–100 μl FACS buffer.
7. Incubate for 20 min on ice.
8. Wash the cells (*see* **Point 5**).
9. Add 50–100 μl SA-PE (or SA-FITC) (5 μg/ml diluted in ice-cold FACS buffer + 1 μg/ml PI). For the PI-only control, add 50–100 μl FACS buffer + 1 μg/ml PI.
10. Incubate for 15 min on ice (protect from light).
11. Wash the cells (*see* **Point 5**).
12. Resuspend the cells in 100 μl FACS buffer and analyze by flow cytometry.

A typical result of a TRAIL receptor surface stain on the leukemic BJAB cell line is shown in **Fig. 2**.

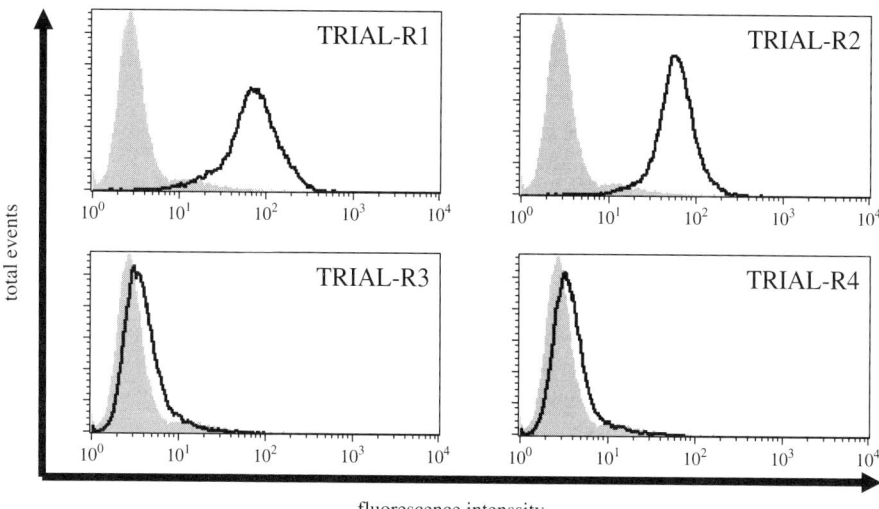

Fig. 2. Flow cytometric analysis of tumor necrosis factor (TNF)-related apoptosis-inducing ligand (TRAIL) receptor surface expression on BJAB cells. Cells were stained with the indicated antibody according to the protocol described, and the fluorescence of propidium iodide (PI) negative cells is shown. The gray area shows the fluorescence of the IgG1-negative control, and the thick line represents the staining seen with antibodies specific for the indicated TRAIL receptors.

3.3. DISC Immunoprecipitation

3.3.1. Prepare Before

1. Cool PBS on ice.
2. Warm RPMI (for most suspension cell lines) or DMEM (for most adherent cell lines) to 37°C.
3. Cool IP lysis buffer on ice and add the COMPLETE protease inhibitor cocktail as recommended by the manufacturer.
4. Prepare 1.5 ml Eppendorf tubes.
5. Prepare stimulation mix (stable for 1 week at 4°C):

    a. DMEM or RPMI with 1 μg/ml FLAG-TRAIL (for TRAIL DISC) + 2 μg/ml M2 antibody and incubate for 30 min at room temperature.
    b. DMEM or RPMI with 0.5 μg/ml FLAG-CD95L (for CD95L DISC) + 1 μg/ml M2 antibody and incubate for 30 min at room temperature.

3.3.2. Stimulation and DISC Formation with Adherent Cells

1. Remove medium from the cells.
2. Wash the cells gently with 12 ml prewarmed DMEM (w/o FBS).

3. Remove medium.
4. Stimulate one half of the cells with 4–6 ml stimulation mix

   a. 5–15 min for FLAG-CD95L in the incubator at 37°C.
   b. 5–30 min for FLAG-TRAIL in the incubator at 37°C.

5. Incubate the control cells for the same time with DMEM (w/o FBS). From now on, all steps have to be carried out on ice with ice-cold buffers (*see* **Note 1**).
6. Stop the stimulation by adding 25 ml ice-cold PBS.
7. Remove the PBS.
8. Rinse the cells again with 25 ml ice-cold PBS.
9. Remove the PBS completely.

3.3.3. Stimulation and DISC Formation with Suspension Cells

1. Spin down 3×10^7 cells at 1200 g for 5 min.
2. Discard the medium.
3. Resuspend the cells gently in 12 ml prewarmed RPMI (w/o FBS).
4. Split the cell suspension into two new 15-ml Falcon tubes.
5. Spin down the cells at 1200 g for 5 min.
6. Discard the medium.
7. Stimulate the cells in one Falcon by resuspending them in 1–2 ml stimulation mix

   a. 5–15 min for FLAG-CD95L in the incubator at 37°C.
   b. 5–30 min for FLAG-TRAIL in the incubator at 37°C.

8. Incubate the control cells for the same time with RPMI (w/o FBS). From now on, all steps have to be carried out on ice with ice-cold buffers (*see* **Note 1**).
9. Stop the stimulation by adding 10 ml ice-cold PBS.
10. Spin down the cells at 1200 g for 5 min.
11. Discard the PBS.
12. Spin down the cells at 1200 g for 5 min.
13. Remove the PBS completely.

3.3.4. Lysis and Immunoprecipitation

1. Add 1 ml IP lysis buffer to the flask (adherent cells) or directly to the cell pellet (suspension cells).
2. Scrape off all the cells with the cell scraper (adherent cells).
3. Transfer the cell suspension into a Eppendorf tube.
4. Lyse the cells by adding 100 μl 10% Triton X100.
5. Incubate on the head to head shaker for 30 min at 4°C.
6. Spin down the debris at 15,000 g in a precooled (4°C) tabletop centrifuge for 30–45 min.
7. Transfer the supernatant into new Eppendorf tubes.
8. Remove 50 μl as pre-IP sample for western blot analysis and freeze away.

9. Add 15–20 μl protein-G beads slurry + 1 μg M2 antibody (anti-FLAG).
10. Add 0.5 μg FLAG-TRAIL (or FLAG-CD95L) to the unstimulated control.
11. Let immunoprecipitate over night on a head to head shaker at 4°C.

3.3.5. Washing the Beads and Preparation of the Samples for SDS–PAGE

Before starting, cool the IP lysis buffer (+1% Triton X100 + COMPLETE protease inhibitor) on ice.

1. Wash the beads:

   a. Spin down the beads in a tabletop centrifuge (850 g, 2 min, 4°C).
   b. Place the tubes on ice for 1–2 min to let the beads settle down.
   c. Remove the supernatant.
   d. Add 1 ml complete IP lysis buffer (+protease inhibitor + 1% Triton X100) per tube.
   e. Mix by inverting gently.

2. Repeat the wash procedure five to six times.
3. Remove the IP lysis buffer completely and add 25 μl 2× LDS protein loading buffer (+50 mM TCEP).
4. Thaw the pre-IP lysates and determine protein concentration by BCA according o the manufacturers' protocol.
5. Adjust the protein concentrations to 20 μg in 20 μl using IP lysis buffer.
6. Add 25 μl 2× LDS protein loading buffer (+50 mM TCEP, add freshly).
7. Incubate the samples for 10 min at 75°C.
8. Cool the samples for 1 min on ice and spin them down at 15,000 g for 1 min on a tabletop centrifuge.

Now the samples are prepared for the separation of the proteins by SDS–PAGE.

3.4. Western Blot Analysis for DISC-Associated Proteins

The recruitment of DISC molecules to the receptor complex can be determined by western blot analysis using specific antibodies for caspase-8 and caspase-10, FADD, the receptor(s), and cFLIP. The pre-IP lysate serves as specificity control for the antibodies used.

3.4.1. Running the Gels

1. Place precast NuPAGE 4–12% gels into the running chamber.
2. Add 400 ml MOPS (3-(N-Morpholino)-propanesulfonic acid) running buffer.
3. Load 15 μl prepared samples.
4. Run gels at 200 V for 55 min.
5. In the meanwhile, prepare nitrocellulose membranes for the transfer (8.5 cm × 7.5 cm).

3.4.2. Transfer

1. Place sponges and membranes in transfer buffer + 10% MeOH.
2. Remove gels from the plates and prepare the blotting sandwich: sponges–filter paper–gel–membrane–filter paper–sponges.
3. Place in transfer box and fill with transfer buffer.
4. Let blot for 2 h at 30 V.

3.4.3. Immunoblot

1. Stain the membranes for 1–3 min with Ponceau red to visualize the proteins.
2. Wash the membranes for 10 min with PBS/0.05% Tween 20 (PBST).
3. Block the membranes in PBST + 5% milk powder for 1–2 h at room temperature.
4. Wash the membranes two times for 5 min in PBST at room temperature.
5. Incubate the membranes in primary antibody solution (1 μg/ml antibody in PBST + 1% milk) over night at 4°C.
6. Wash the blots five times for 5 min with PBST.
7. Incubate with secondary antibody 1:20,000 diluted in PBST + 1% milk for 1 h at room temperature.

3.4.4. Developing

1. Wash the membranes five times for 5 min with PBST.
2. Remove the fluid and add ECL solution; make sure that the membrane is fully covered.
3. Incubate for 1 min at room temperature.
4. Place the membrane between transparent sheets.
5. Expose to film under safety light conditions and develop the film.

3.4.5. Stripping and Reprobing

After detection with the first antibody, the HRP-conjugated secondary antibody can be stripped off by a low pH buffer, and the membranes can be used again for detection of a new protein. This gentle stripping procedure mainly destroys the HRP conjugated to the secondary antibody.

1. Wash the membranes two times for 5 min in PBST.
2. Incubate the membranes for 20 min in stripping buffer at room temperature.
3. Wash the membranes two times for 5 min in PBST.
4. Continue with blocking the membranes and adding the next primary antibody.
5. This could be repeated maximal for three to four times.

A typical result of a biochemical TRAIL DISC analysis with the B-lymphoma cell line (BJAB) is shown in **Fig. 3**.

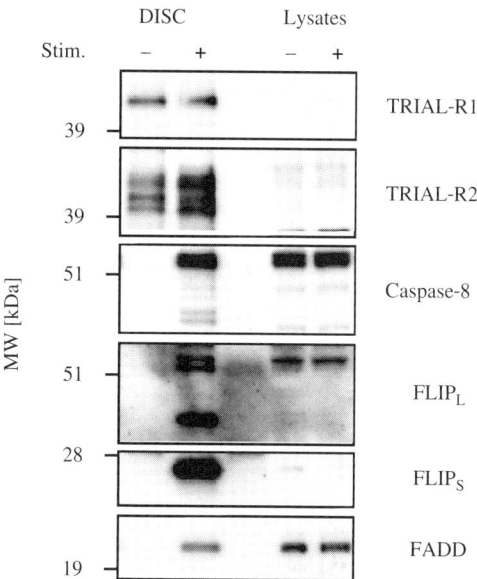

Fig. 3. Biochemical analysis of the tumor necrosis factor (TNF)-related apoptosis-inducing ligand (TRAIL) death-inducing signaling complex (DISC) in BJAB cells. Isolated receptor complexes were separated by SDS–PAGE and immunoblotted for the proteins indicated. After detection, the membranes were stripped and probed with an antibody for the next protein.

4. Notes

1. It is very important to perform the entire procedure of lysis and immunoprecipitation at low temperature (4°C or on ice) to keep the complex stable.
2. Some members of the TNFSF do not bind their cognate receptor when the cells are lysed before adding the ligand (e.g., CD95L–CD95). This has to be taken in account for the unstimulated controls.
3. The stripping procedure does not remove the primary antibody, thus it is well possible that it still signals if the subsequent antibody has the same isotype.
4. The anti-FLAG (M2) antibody is a mouse IgG1 isotype and may interfere with the detection of FLIP as the NF6 anti-FLIP antibody is also a mouse IgG1 antibody and the molecular weight of cFLIP$_L$ is very similar to the molecular weight of the M2 heavy chain.

References

1. Walczak, H., Miller, R. E., Ariail, K., Gliniak, B., Griffith, T. S., Kubin, M., Chin, W., Jones, J., Woodward, A., Le, T., Smith, C., Smolak, P., Goodwin, R. G., Rauch, C. T., Schuh, J. C., and Lynch, D. H. (1999) *Nat Med* **5**, 157–63.

2. Kelley, S. K., and Ashkenazi, A. (2004) *Curr Opin Pharmacol* **4,** 333–9.
3. Wiley, S. R., Schooley, K., Smolak, P. J., Din, W. S., Huang, C. P., Nicholl, J. K., Sutherland, G. R., Smith, T. D., Rauch, C., Smith, C. A., and et al. (1995) *Immunity* **3,** 673–82.
4. Kim, Y., and Seol, D. W. (2003) *Mol Cells* **15,** 283–93.
5. LeBlanc, H. N., and Ashkenazi, A. (2003) *Cell Death Differ* **10,** 66–75.
6. Pan, G., O'Rourke, K., Chinnaiyan, A. M., Gentz, R., Ebner, R., Ni, J., and Dixit, V. M. (1997) *Science* **276,** 111–3.
7. Walczak, H., Degli-Esposti, M. A., Johnson, R. S., Smolak, P. J., Waugh, J. Y., Boiani, N., Timour, M. S., Gerhart, M. J., Schooley, K. A., Smith, C. A., Goodwin, R. G., and Rauch, C. T. (1997) *EMBO J* **16,** 5386–97.
8. Somasundaram, K., Zhang, H., Zeng, Y. X., Houvras, Y., Peng, Y., Wu, G. S., Licht, J. D., Weber, B. L., and El-Deiry, W. S. (1997) *Nature* **389,** 187–90.
9. Chaudhary, P. M., Eby, M., Jasmin, A., Bookwalter, A., Murray, J., and Hood, L. (1997) *Immunity* **7,** 821–30.
10. Schneider, P., Bodmer, J. L., Thome, M., Hofmann, K., Holler, N., and Tschopp, J. (1997) *FEBS Lett* **416,** 329–34.
11. Screaton, G. R., Mongkolsapaya, J., Xu, X. N., Cowper, A. E., McMichael, A. J., and Bell, J. I. (1997) *Curr Biol* **7,** 693–6.
12. Pan, G., Ni, J., Wei, Y. F., Yu, G., Gentz, R., and Dixit, V. M. (1997) *Science* **277,** 815–8.
13. Sheridan, J. P., Marsters, S. A., Pitti, R. M., Gurney, A., Skubatch, M., Baldwin, D., Ramakrishnan, L., Gray, C. L., Baker, K., Wood, W. I., Goddard, A. D., Godowski, P., and Ashkenazi, A. (1997) *Science* **277,** 818–21.
14. Kimberley, F. C., and Screaton, G. R. (2004) *Cell Res* **14,** 359–72.
15. Degli-Esposti, M. (1999) *J Leukoc Biol* **65,** 535–42.
16. Bodmer, J. L., Holler, N., Reynard, S., Vinciguerra, P., Schneider, P., Juo, P., Blenis, J., and Tschopp, J. (2000) *Nat Cell Biol* **2,** 241–3.
17. Kischkel, F. C., Lawrence, D. A., Chuntharapai, A., Schow, P., Kim, K. J., and Ashkenazi, A. (2000) *Immunity* **12,** 611–20.
18. Sprick, M. R., Weigand, M. A., Rieser, E., Rauch, C. T., Juo, P., Blenis, J., Krammer, P. H., and Walczak, H. (2000) *Immunity* **12,** 599–609.
19. Kischkel, F. C., Lawrence, D. A., Tinel, A., LeBlanc, H., Virmani, A., Schow, P., Gazdar, A., Blenis, J., Arnott, D., and Ashkenazi, A. (2001) *J Biol Chem* **276,** 46639–46.
20. Sprick, M. R., Rieser, E., Stahl, H., Grosse-Wilde, A., Weigand, M. A., and Walczak, H. (2002) *EMBO J* **21,** 4520–30.
21. Thorburn, A. (2004) *Cell Signal* **16,** 139–44.
22. Wang, J., Chun, H. J., Wong, W., Spencer, D. M., and Lenardo, M. J. (2001) *Proc Natl Acad Sci USA* **98,** 13884–8.
23. Wang, J., Zheng, L., Lobito, A., Chan, F. K., Dale, J., Sneller, M., Yao, X., Puck, J. M., Straus, S. E., and Lenardo, M. J. (1999) *Cell* **98,** 47–58.

24. Golks, A., Brenner, D., Fritsch, C., Krammer, P. H., and Lavrik, I. N. (2005) *J Biol Chem* **280,** 14507–13.
25. Krueger, A., Schmitz, I., Baumann, S., Krammer, P. H., and Kirchhoff, S. (2001) *J Biol Chem* **276,** 20633–40.
26. Ganten, T. M., Haas, T. L., Sykora, J., Stahl, H., Sprick, M. R., Fas, S. C., Krueger, A., Weigand, M. A., Grosse-Wilde, A., Stremmel, W., Krammer, P. H., and Walczak, H. (2004) *Cell Death Differ* **11 Suppl 1,** S86–96.
27. Sharp, D. A., Lawrence, D. A., and Ashkenazi, A. (2005) *J Biol Chem* **280,** 19401–9.
28. Zhang, L., and Fang, B. (2005) *Cancer Gene Ther* **12,** 228–37.
29. Peter, M. E. (2004) *Biochem J* **382,** e1–3.
30. Xiao, C., Yang, B. F., Asadi, N., Beguinot, F., and Hao, C. (2002) *J Biol Chem* **277,** 25020–5.
31. Ricci-Vitiani, L., Pedini, F., Mollinari, C., Condorelli, G., Bonci, D., Bez, A., Colombo, A., Parati, E., Peschle, C., and De Maria, R. (2004) *J Exp Med* **200,** 1257–66.

17

Laser Microdissection Sample Preparation for RNA Analyses

Christopher J. Vega

Summary

Gene expression analysis provides an insight into the unique and defining biomolecular characteristics of a given cell type. However, heterogeneous cellular compositions hinder gene analysis studies from most tissue samples. The laser microdissection (LMD) technique allows for the unambiguous isolation of a desired cell population. However, preserving RNA integrity can be challenging because of the deliberately limited amount of starting material, sometimes as little as a single cell. General laboratory procedures for reducing ribonuclease (RNase) activity, both in reagents and in the laboratory environment, are required for successful downstream RNA isolation and quantitation. Quality RNA can be extracted from sections made from flash-frozen and paraffin-embedded tissue. The standard histological stains such as hematoxylin and eosin (H&E), or toluidine blue, can provide visualization of the cells of interest. Following LMD, validation of RNA integrity should precede downstream analysis.

Key Words: Laser microdissection; laser capture microdissection; gene expression; RNA; sample preparation; RNase.

1. Introduction

Laser microdissection (LMD), or laser capture microdissection (LCM), is a technique for isolating microscopic samples of interest. The benefit of this technique, with regard to studies of gene expression within heterogeneous tissues, is that the cells of interest can be studied independently of their surrounding. That is, the unique expression profile of the targeted cells will not be obscured by expression levels contributed from neighboring cells. However, the neighboring cells are not without value as they can be captured separately to allow comparative studies, for example, expression variations in cancerous tissue versus non-cancerous tissue.

From: *Methods in Molecular Biology, vol. 414: Apoptosis and Cancer*
Edited by: G. Mor and A. B. Alvero © Humana Press Inc., Totowa, NJ

1.1. Suppressing Ribonuclease Activity

Handling and preparing samples, to protect RNA from degradation, is paramount to the success of LMD isolation for gene expression analyses. The ribonuclease (RNase) family of enzymes catalyzes the cleavage of nucleotides in RNA leading to degradation. Unfortunately, RNases are everywhere. The ubiquitous nature of these molecules makes working with tissue, for the purpose of isolating RNA, a challenging endeavor.

Methods for reducing the effects of RNases within solutions and upon laboratory equipment exist, but RNase enzymes are resilient. Bearing this in mind, RNases will not be eliminated; they will be either denatured or chemically modified, or both, to reduce enzymatic activity.

When performing RNA experiments, it is best to dedicate reserved locations, reagents, and equipment. Always wear clean gloves when handling any materials (sample, reagents, equipment, etc.) intended for work with RNA. These steps will aid in reducing potential RNase contamination and subsequent sample degradation.

1.1.1. Suppressing RNase Activity in Experimental Reagents with Diethyl Pyrocarbonate

Autoclaving alone may not be effective in eliminating RNases; therefore, it is common practice to chemically treat solutions and containers for use with RNA. Diethyl pyrocarbonate (DEPC) is commonly used for this purpose, with the exception of TRIS buffers, as DEPC reacts directly with TRIS. DEPC covalently modifies the histidine residues of RNases and renders the enzyme inactive. But be aware that DEPC will also react with RNA, so DEPC-treated solutions must be heated to break down DEPC to CO_2 and ethanol.

The protocol for DEPC treatment is uncomplicated, however, DEPC is a toxic substance and suspected carcinogen. As such, care should be taken with its use.

1.1.2. Suppressing RNase Activity on Laboratory Equipment

RNaseZap is a commercial product (Ambion, Austin, TX) commonly used to treat the surfaces of glassware, plastic surfaces, countertops, and pipettes to reduce RNase contamination in the laboratory. RNaseZap can be applied directly to the surfaces of laboratory items (although it is not recommended for metal surfaces that may corrode). The solution works on contact, and the treated equipment is ready for use following rinsing with distilled H_2O and drying. As always, wear gloves, which will then be RNaseZap-treated as well.

1.2. Tissue Sectioning

Tissue sections compatible with the LMD method are prepared by one of the two methods, cryostat sectioning of flash-frozen tissue or microtome sectioning of fixed, paraffin-embedded tissues. Each method has its benefits and drawbacks. Samples for cryostat sectioning can be prepared quickly with flash-freezing, however, tissue morphology noticeably suffers. Paraffin embedding is time consuming, but the tissue morphology from the resulting sections is superior to those from a cryostat. However, samples for paraffin embedding must be prepared with a fixative solution (typically formalin) prior to embedding. Formaldehyde-based fixatives will negatively impact RNA extraction and quality. Alternatively, alcohol-based fixatives are available that have better compatibility for work with RNA.

Whether working with frozen or paraffin-embedded tissues, every precaution should be taken to maintain the sectioning equipment and environment in an RNase-free state. Disposable blades for sectioning should be used, if possible, to reduce RNases but also to provide the best consistency in sections. A good starting thickness for sections is 10 µm. This can be varied depending on sample and application.

The Leica LMD6000 and its predecessor, the Leica AS LMD, use slides that have a polyethylene (POL)-based, UV-absorbing membrane. The most commonly used slide for RNA work is the polyethylene naphthalate (PEN)-coated glass slide. These slides can be handled like standard glass slides with tissue sections placed on the membrane side of the slides.

Alternatively, there are slides available with POL or polyethylene terephthalate (PET) membranes extended over a metal frame. There is a small cavity on one side of the slide because of the thickness of the metal frame, and the section is mounted on the membrane surface on the backside of the cavity. As there is no backing material to work as a heat sync when transferring cryosections, Leica offers a plexiglass support frame that fits into this cavity to aid thermal section transfer.

1.2.1. Preparation of Cryostat Tissue Sections

Before sectioning, an embedding compound, for example, OCT, is required to adhere the tissue to the microtome's specimen holder. This compound will solidify when frozen. However, excess embedding compound can interfere with cutting during LMD, so its use should be done so sparingly. Preparations made from disposable molds, in which samples are surrounded by embedding compound and then frozen, should have the excess embedding compound trimmed away to a minimum. OCT is water soluble, so if the tissue is to be immediately sectioned, without additional processing, then slide mounted sections should be washed in 70% ethanol to clear remaining OCT.

The tissue can be flash-frozen either before or after being placed in the embedding compound. The benefit of freezing before embedding is that it is easier to gauge the progress of the tissue freezing. However, when placing frozen tissue in embedding compound, the tissue will thaw to some degree. The benefit of embedding first, then freezing is to ensure a single freezing process.

The cryostat should be at the appropriate temperature for the tissue type to be sectioned before the tissue block and specimen holder are mounted in the microtome. Cryostat chamber temperatures in the range of –15 and –30°C are common for flash-frozen tissues. Tissue sections can be made at thickness between 5 and 15 μm, but this can be optimized based on the tissue and application.

Before performing any staining protocols, the tissue sections should be fixed with 70% ethanol or ice-cold acetone for approximately 30 s.

1.2.2. Preparation of Paraffin-Embedded Tissue Sections

Paraffin embedding requires more processing of the tissue than the flash-freezing process. The tissue must be preserved, that is, fixed, typically with formalin. However, formaldehyde-based fixatives cross-link and degrade RNA. Alcohol-based fixatives, such as Carnoy's solution or methacarn, are non-crosslinking, protein-precipitating fixatives (1) and have been successfully used as a fixative for paraffin-embedded tissues for downstream RNA analysis from LMD-isolated samples (2,3).

Following adequate fixation, the tissue must be run through a series of solutions that will take it from an aqueous state and allow it to be infiltrated by a non-aqueous support medium, typically paraffin wax. The process begins by dehydrating the tissue through a series of graded ethanol solutions. The tissue is then exposed to a "clearant" that is an organic solvent, such as xylene, that is miscible with both alcohol and paraffin.

As all the processes mentioned above require diffusion of solutions into the tissue, it is recommended that the tissue not be larger than 5 mm in any dimension. Following infusion of paraffin into the tissue, extra paraffin should be trimmed before microtome cutting.

Sections of the paraffin-embedded tissues should be obtained using a rotary microtome. Tissue sections with thickness between 5 and 15 μm should be made into ribbons and stretched on a slide warmer prior to mounting on LMD slides.

Before performing any staining process on the slide-mounted tissue sections, the tissue must be free of paraffin. This is easily done with organic solvents and a series of graded alcohol washes.

1.3. Tissue Section Staining

To minimize RNase activity and provide sufficient detail from the staining process, a balance between the staining time and RNase activity must be observed. The staining times therefore should be optimized for the tissue, section thickness, and stain. A rough rule of thumb is that the total time for the staining procedure should not exceed 30 min, and once stained, the sample should be used immediately for LMD.

There are many common histological staining protocols, but two are presented in this section for simplicity. The first, hematoxylin and eosin (H&E), renders nuclei and some mitochondria blue and cytoplasmic compartments pink. The second, toluidine blue, renders nuclei blue to purple.

1.4. LMD

During the LMD process, it is best to have a sample that has been fully desiccated. This will ensure the best cutting and lowest RNase activity.

The local environment of the instrument is also a matter of concern. For example, relative humidity should be in the range of 30–40% for the best cutting results. Excessive relative humidity will encourage RNase activity and has adverse effects on laboratory equipment. Low levels, below 30%, may allow static charges to affect sample collection. A humidity monitor (hygrometer) can be purchased at most home center locations and is recommended for determining the room's moisture level. The area's humidity level can be regulated with a consumer grade humidifier or dehumidifier depending on conditions.

If static charge issues persist, a device known as the "ZeroStat" gun (Armour Home Electronics, Hertfordshire, UK) can be used to attempt to neutralize the static charge, at the level of the sample.

2. Materials
2.1. Suppressing RNase Activity
2.1.1. Equipment

1. Stir-plate.
2. Autoclave.
3. Autoclavable glassware.

2.1.2. Reagents

1. DEPC (cat. no. D5758, Sigma–Aldrich, St. Louis, MO, USA).
2. RNaseZap (Ambion).

2.2. Tissue Sectioning

Whether tissue sections are produced on a cryostat or rotary microtome, LMD compatible slides are required, and these include Leica's PEN, PET, or POL LMD slides.

2.2.1. Preparation of Cryostat Tissue Sections

2.2.1.1. EQUIPMENT

1. Beaker.
2. Bucket for dry ice.
3. Leica CM3050 S cryostat.

2.2.1.2. REAGENTS

1. OCT embedding compound.
2. Dry ice.
3. Liquid nitrogen.
4. Isopentane.

2.2.2. Preparation of Paraffin-Embedded Tissue Sections

2.2.2.1. EQUIPMENT

1. Paraffin-embedding molds.
2. Leica EG1150 H paraffin-embedding station.
3. Leica RM2255 microtome.

2.2.2.2. REAGENTS

1. Alcohol-based fixative (solutions list by v/v)

   a. Carnoy's fixative: 60% absolute ethanol, 30% chloroform, and 10% glaciec acetic acid.
   b. Methacarn: 60% absolute methanol, 30% chloroform, and 10% glaciec acetic acid.

2. Graded ethanol series: 70% ethanol, 95% ethanol, and 100% ethanol.
3. Xylene, reagent grade.
4. Paraffin wax.

2.3. Tissue Section Staining

2.3.1. Toluidine Blue

2.3.1.1. EQUIPMENT

1. 0.22-μm sterile syringe filter.
2. Slide rack and staining reservoirs.

2.3.1.2. REAGENTS

1. Toluidine blue.
2. DEPC-treated water.
3. 75% ethanol.

2.3.2. H&E

2.3.2.1. EQUIPMENT

1. Slide rack and staining reservoirs.

2.3.2.2. REAGENTS

1. Ethanol series: 70, 95, and 100%.
2. Hematoxylin solution, Harris-modified (cat. no. HHS-16, Sigma–Aldrich, St. Louis, MO, USA).
3. Eosin Y solution, alcoholic, with phloxine (cat. no. HT110-3-32, Sigma–Aldrich, St. Louis, MO, USA).
4. Bluing reagent 0.1% NH_4OH (cat. no. 443182, Sigma–Aldrich, St. Louis, MO, USA) in H_2O.
5. Distilled H_2O.
6. DEPC-treated water.

2.4. LMD

2.4.1. Equipment

1. Leica LMD6000.
2. 0.2- or 0.5-ml PCR tubes for sample collection.
3. Dehumidifier or humidifier.
4. ZeroStat gun.

2.4.2. Reagents

1. Lysis buffer (from RNA extraction kit).

3. Methods

3.1. Suppressing RNase Activity

3.1.1. Suppressing RNase Activity in Experimental Reagents with DEPC

1. Prepare all solutions under a fume hood.
2. Wear gloves and protective eyewear.
3. Use only thoroughly cleaned glassware.
4. Open stock DEPC container slowly as pressure may have built up within the bottle.
5. Make solutions to DEPC concentration of 0.1% v/v.
6. Stir vigorously to ensure DEPC is thoroughly mixed within the solution.
7. Treat the solution for a minimum of 12 h (typically this is done overnight).
8. Autoclave for a minimum of 15 min for the decomposition of DEPC.

3.2. Tissue Sectioning

Depending on the materials and equipment available in the laboratory, four flash-freezing options are presented.

3.2.1. Preparation of Cryostat Tissue Sections

3.2.1.1. Isopentane and Dry Ice (Working Under Fume Hood)

1. Place crushed, or powdered, dry ice in ice bucket.
2. Make a hollow in the ice in the center of bucket for a beaker.
3. Add enough isopentane to a pyrex beaker to submerge tissue and allow for 5–10 min to cool.
4. Add tissue (which should turn bright white) to isopentane.
5. Remove tissue from isopentane and tap tissue. If it sounds dull or hollow, then it is not frozen. Continue until it sounds solid.

3.2.1.2. Isopentane and Dry Ice Slurry (Working Under Fume Hood)

1. Pre-cool 2-methyl-butane in a beaker surrounded by dry ice. This will help the 2-methyl-butane from bubbling over when the dry ice is added.
2. In a beaker or specimen container, add crushed dry ice to the 2-methyl-butane to make a slurry mixture of the two.
3. When bubbling stops, fix the 2-methyl-butane at correct freezing temperature.
4. Immerse the tissue slowly. Eventually, it will sink to the bottom of the 2-methyl-butane.

3.2.1.3. Isopentane and Liquid Nitrogen (Working Under Fume Hood)

1. Pre-cool 2-methyl-butane in a beaker surrounded by liquid nitrogen.
2. Allow 2-methyl-butane to cool for 5–10 min.

3. Add tissue (which should turn bright white when frozen) to isopentane.
4. Remove tissue from isopentane and tap tissue. If it sounds dull or hollow, then it is not frozen. Continue until it sounds solid.

3.2.1.4. FLASH-FREEZING IN LIQUID NITROGEN

1. Place liquid nitrogen in a Styrofoam® container.
2. Place the Styrofoam® container inside a Petri dish lid (a support rack to hold the Petri dish lid may be needed).
3. Place the tissue in a disposable mold and embed in the tissue-freezing medium, or alternatively, place tissue embedded in the tissue-freezing medium on a coverslip and place it in the liquid nitrogen.

3.2.2. Preparation of Paraffin-Embedded Tissue Sections

3.2.2.1. FIXATION

Carnoy's solution and methacarn should be freshly prepared before tissue fixation and stored at 4°C until use. Small tissue samples should be submerged in 5–10 times their volume in fixation for 2–12 h (depending on tissue size) with gentle agitation at 4°C.

3.2.2.2. DEHYDRATION AND CLEARING PROCESS

Transfer the tissue to a graded series of alcohol and then to the clearant. During the dehydration and clearing steps, the wax should be melted to the appropriate temperature, typically 60°C.

1. 70% ethanol for 45 min.
2. 95% ethanol for 45 min.
3. 100% ethanol for 45 min.
4. 100% ethanol for 45 min.
5. Xylene for 45 min.
6. Xylene for 45 min.

3.2.2.3. PARAFFIN INFILTRATION PROCESS

The paraffin infiltration process is completed by exposing the specimen to several changes of molten paraffin and can be accelerated by working in a vacuum. The final change of paraffin must be clean and cannot contain residual clearant. Once infiltration is complete, the tissue can be placed face down in a mold containing clean, melted paraffin, and then cooled to create a block for sectioning.

3.3. Tissue Section Staining

3.3.1. Toluidine Blue

To make a working 1% toluidine blue solution, dissolve 0.1 g toluidine blue in 10 ml DEPC-treated water and pass through a sterile filter.

1. Toluidine blue solution for 3 min.
2. DEPC water for 15 s.
3. DEPC water for 15 s.
4. 75% ethanol for 3 min.
5. Air dry.

3.3.2. H&E

1. Distilled H_2O for 30 s.
2. Hematoxylin for 1 min.
3. DEPC-treated water for 30 s.
4. Bluing reagent for 30 s.
5. Eosin for 10 s.
6. 70% ethanol for 30 s.
7. 95% ethanol for 30 s.
8. 100% ethanol for 30 s.
9. Air dry.

3.4. LMD

Turn the LMD system on (computer, microscope control box, and laser unit). Start the Leica Laser Microdissection software. Place a slide, specimen facing down, in a holder and place in stage insert.

Place a PCR tube, cap first and tube folded underneath, in the collection device. Add a small amount of lysis buffer to the tube cap and load the collection device into the microdissection stage. Each cap is uniquely identified in the software and can be targeted for sample collection in the software's tube cap interface.

Regions of interest, that is, cells to be collected, are identified by outlining their shapes and can be completed using two different modes within the software. Regions can be identified one at a time by selecting "Single Shape" so that only one region is drawn within the software. By choosing "Multiple Shapes," several regions of interest can be selected; additionally, these regions can be marked and collected into different wells of the collection device.

The software also allows two drawing modes: "Draw + Cut" or "Move + Cut." With "Draw + Cut," either single or multiple shapes are created with drawing tools: line, circle, and rectangle. The circle and rectangle shapes automatically generate a "closed shape," that is, there are no gaps in the laser cut line. Use of the line tool, which is a free-hand drawing tool for complex shapes, will not generate a "closed shape" unless the "close line(s)" option is checked. To collect the regions of interest, select the "Start Cut" button. The laser then cuts along the perimeter of all the shapes specified. If the "Move + Cut" option is selected, the laser will cut "live" in a free-hand fashion, which has been described as a "laser-scalpel" manner of cutting.

When the system cuts, a small ablation path results as the laser follows the course of the shapes drawn. This ablation path must be taken into consideration when drawing shapes. Additionally, rounded shapes tend to release from the slide more easily than points or corners.

For users that require high-throughput cell collection, the optional Auto Vision Control (AVC) module (accessory to the LMD software) performs a cell-recognition algorithm to identify desired cells based on criteria such as color and size.

To ensure collection and that regions of interest have been collected, microdissectate can be inspected in the caps from within the software. Once the desired quantity of cells has been collected, the collection device can be unloaded from the LMD system, and the PCR tube with the cells of interest can be used to complete downstream RNA methods.

4. Notes

4.1. Tissue Time

While in tissue, RNA will be vulnerable to degradation. The goal of LMD sample preparation is to reduce the time required to take the sample of interest from living tissue into lysis buffer, that is, tissue time. To elaborate, lengthy staining procedures that increase tissue time will very likely have an adverse effect on RNA quality. This is always a compromise; the better the morphology, the greater the sacrifice of molecules of interest, that is, RNA. As such, one will benefit from a faster, albeit lighter staining, when downstream RNA analyses are performed. To reduce tissue time, many users work toward a target of 30 min to take the cells of interest from tissue to lysis buffer.

4.2. How Many Cells to Collect?

A starting point for the novice would be hundreds of cells. The number of cells needed for analysis depends on a variety of criteria such as tissue type, abundance of mRNA of interest, extraction kit, and mRNA quantitation method, among others. So the exact number of starting cells varies.

As a frame of reference, the number of cells a given LMD-based experimenter will collect can be divided into three groups, one-third collect more than 1000 cells, one-third collect between 100 and 1000 cells, and one-third collects less than 100, even down to a single cell.

4.3. How Much RNA will I Obtain?

As the amount of starting material changes, so should the kit used for RNA isolation. Kits designed for different amounts of starting material will concentrate the extracted RNA into a range that will be acceptable for downstream analysis protocols, for example, real-time PCR.

The total amount of RNA in a mammalian cell is typically on the order of 10–30 pg. As a rough rule of thumb, 100 cells will provide 1 ng total RNA. This relationship can be extended linearly so that 10 cells would result in 0.1 ng or 1000 cells would provide 10 ng, and so on. Keep in mind, however, that the proportion of mRNA will only be 1–3% of the total RNA amount.

4.4. Determine the Quality of Extracted RNA

Before pursuing further downstream analysis, determine the quality of the purified RNA. Chip-based, electrophoretic systems, for example, Agilent Bioanalyzer, can be used to assess RNA quality with a minimum of extracted RNA, using only 1 μl per "lane." The system can analyze the peak shoulders, or elevation of baseline, of 28s and 18s ribosomal RNA present within the sample that indicates RNA degradation.

Acknowledgments

The author thank Bob Fasulka and Andy Lee, Laser Microdissection Application Specialists, Leica Microsystems, for sharing their technical expertise and reviewing this document. Additional thanks are extended to Jan Minshew, HT, HTL (ASCP, Marketing Manager, Leica Microsystems) for reviewing the protocols on sample preparation, and Pam Jandura, Marketing Specialist, Leica Microsystems, for editorial assistance.

References

1. Puchtler, H., et al., *Histochemie* 1970; 21(2): 97–116.
2. Takagi, H., et al., *J. Histochem. Cytochem.* 2004; 52(7): 903–13.
3. Jiang, K., et al., *Plant Mol. Biol.* 2006; 60(3): 343–63.

Index